BASIC BUSINESS MATH

Robert L. Dansby, Ph.D.
Columbus Technical Institute

PRENTICE HALL Englewood Cliffs, New Jersey 07632

Library of Congress Cataloging-in-Publication Data
Dansby, Robert L.
Basic business math / Robert L. Dansby.
p. cm.
ISBN 0-13-060971-4
1. Business mathematics—Problems, exercises, etc. I. Title.
HF5691.D36 1992 91-37186
650'.01'513—dc20 CIP

DEDICATION

This book is dedicated to my wife, Barbara; and to my son, Champ—for without them, no project would be meaningful.

Acquisition editor: James C. Boyd
Production editor: Jacqueline A. Martin
Copy editor: Karen Verde
Designer: Linda J. Den Heyer Rosa
Cover designer: Rosemarie Paccione
Prepress buyer: Ilene Levy
Manufacturing buyer: Ed O'Dougherty
Supplements editor: Lisamarie Brassini / Cindy Harford
Editorial assistant: Irene Hess

A Simon & Schuster Company
Englewood Cliffs, New Jersey 07632

Printed in the United States of America
10 9 8 7 6 5 4 3 2 1

ISBN 0-13-060971-4

Prentice-Hall International (UK) Limited, *London*
Prentice-Hall of Australia Pty. Limited, *Sydney*
Prentice-Hall Canada Inc., *Toronto*
Prentice-Hall Hispanoamericana, S.A., *Mexico*
Prentice-Hall of India Private Limited, *New Delhi*
Prentice-Hall of Japan, Inc., *Tokyo*
Simon & Schuster Asia Pte. Ltd., *Singapore*
Editora Prentice-Hall do Brasil, Ltda., *Rio de Janeiro*

Contents

Preface v

SECTION I **Whole Numbers**
UNIT 1 Reading and Writing Whole Numbers 1
UNIT 2 Rounding Off Whole Numbers 5

SECTION II **Addition of Whole Numbers**
UNIT 3 Adding Whole Numbers 11
UNIT 4 Combinations That Total Ten 21
UNIT 5 Horizontal Addition 27

SECTION III **Subtraction of Whole Numbers**
UNIT 6 Subtracting Whole Numbers 37
UNIT 7 Subtracting When Borrowing Is Needed 45

SECTION IV **Multiplication of Whole Numbers**
UNIT 8 Multiplying Whole Numbers 53
UNIT 9 Multiplying Numbers with Two or More Digits in the Multiplier 61
UNIT 10 Shortcut Multiplication 71

SECTION V **Division of Whole Numbers**
UNIT 11 Dividing Whole Numbers 81
UNIT 12 Division with a Remainder and Finding Averages 91
UNIT 13 Dividing When Zeros are Needed in the Quotient 101
UNIT 14 Shortcut Division 111

SECTION VI **Fractions**
UNIT 15 The Meaning of Fractions 117
UNIT 16 Reducing Fractions to Lowest Terms 123
UNIT 17 Determining the Greatest Common Divisor 129
UNIT 18 Raising Fractions to Higher Terms (Higher Equivalents) 135
UNIT 19 Improper Fractions and Mixed Numbers 141

SECTION VII **Addition and Subtraction of Fractions**
UNIT 20 Adding Fractions with the Same Denominator 147
UNIT 21 Adding Fractions with Different Denominators 153
UNIT 22 Adding Mixed Numbers 163

UNIT 23 Subtracting Fractions 173
UNIT 24 Subtracting Mixed Numbers 181

SECTION VIII Multiplication of Fractions
UNIT 25 Multiplying Fractions 191
UNIT 26 Multiplying Mixed Numbers 199
UNIT 27 Multiplying Large Mixed Numbers 209

SECTION IX Division of Fractions
UNIT 28 Dividing Fractions 219
UNIT 29 Dividing Mixed Numbers 225

SECTION X Decimal Fractions
UNIT 30 The Meaning of Decimals and Reading Decimals 233
UNIT 31 Rounding Off Decimals 243
UNIT 32 Comparing Decimals 251

SECTION XI Addition, Subtraction, and Multiplication of Decimals
UNIT 33 Adding and Subtracting Decimals and Mixed Numbers 257
UNIT 34 Multiplying Decimals 273
UNIT 35 Multiplying Decimals Using Shortcuts 285

SECTION XII Division of Decimals
UNIT 36 Dividing Decimals 293
UNIT 37 Dividing Decimals Using Shortcuts 307
UNIT 38 Equivalent Decimal and Common Fractions 315

SECTION XIII Percents and Aliquot Parts
UNIT 39 Understanding Percents 327
UNIT 40 Working with the Base, Rate, and Percentage 337
UNIT 41 Additional Percent Applications 349
UNIT 42 Aliquot Parts 361

SECTION XIV Business Application of Percents
UNIT 43 Trade Discounts 371
UNIT 44 Two or More Trade Discounts in a Series 377
UNIT 45 Cash Discounts 389
UNIT 46 Computing Markup and Selling Price 397
UNIT 47 Computing Markdown 409
UNIT 48 Sales Taxes 415
UNIT 49 Finding Commissions 421
UNIT 50 Interest 429

SECTION XV Earnings
UNIT 51 Determining Gross Pay 435
UNIT 52 Payroll Deductions and the Payroll Register 443

SECTION XVI Banking
UNIT 53 Bank Checking Account 455
UNIT 54 Bank Reconciliation 465
Appendix A Working with Measurements A-1
Appendix B Compound Interest B-1

Preface

We live in a fast-moving, rapidly changing world. Skills in basic business mathematics are needed to enable you to begin your career and move up in that career. According to the latest research, the average professional person (bookkeeper, accountant, office worker, manager, etc.) will change positions at least six times during a working career. Many of these changes will take place because individuals will have the educational and occupational skills needed to keep moving up the success ladder.

Basic Business Math is designed to give you the basic math skills needed to enter the work force. It is a very user-friendly text that is aimed at the following objectives:

1. Thoroughly reviewing the fundamental operations of arithmetic: whole numbers (addition, subtraction, multiplication, and division), fractions, decimals, and percents
2. Developing the student's ability to apply the basic math skills to common business situations
3. Developing the student's ability to solve common business problems involving discounts, payroll, interest, markup, and banking
4. Developing the student's ability to use shortcuts and work mentally with speed and accuracy

Basic Business Math has been designed to allow you to work on an individual basis or receive instruction in the traditional classroom setting. It teaches by introducing concepts, supported by examples, and immediate feedback. Text matter and problems move from easy to more challenging.

Study your business math concepts well. You will be gaining skills that will benefit you throughout your entire working career as well as in your personal life.

ACKNOWLEDGMENTS

I would like to express my sincere thanks to my many academic colleagues who offered many excellent suggestions and recommendations during the review process of this book and its supplements. This project would not have been possible without their expert assistance. Their time, experience, and expertise are greatly appreciated.

John Bloomquist	Lamson Junior College, Phoenix, AZ
Angelo M. Bonoan, Jr.	Union City SCS Technical Institute, Union City, NJ
Richard P. Burgholzer	Rochester Business Institute, Rochester, NY
Lerty B. Burnette	Miami Lakes Technical Education Center, Miami Lakes, FL
Pauline Capece	PSI Institute of Philadelphia (Programming Systems Incorporated), Philadelphia, PA
Marian Cargile	Balin Institute of Technology, Oceanside, CA
Katherine King	NEC (National Education Center), Little Rock, AR
Dennis McCollum	Columbus Technical Institute, Columbus, GA
Nita McKenzie	Singer Business College, North Hollywood, CA
William J. Morris	Northwestern Business College, Chicago, IL
Donna Norton	Barclay College, Sacremento, CA
James A. Page	ATI (American Trades Institute), Dallas, TX
Weston L. Perry-Fiske	Lamson Junior College, Phoenix, AZ
Rosemary Piserchio	College of San Mateo, San Mateo, CA
Lance A. Shaw	Blake Business School, New York, NY

Special thanks are also in order to my good friends at Prentice Hall, who are responsible for this book. My sincere appreciation goes to Jim Boyd, Editor-in-Chief; Bob Kern, Marketing Manager; and Lisamarie Brassini and Cindy Harford, Supplements Editors.

Accolades are also in order for my Production Editor, Jackie Martin, for her enthusiasm, promptness, and unflagging devotion to the project.

Finally, I would like to recognize the significant contribution of my good friend and professional colleague, Dennis McCollum. Dennis read every page of the manuscript and checked every answer in the book. His professional help and moral support are immensely appreciated.

UNIT 1 Reading and Writing Whole Numbers

A **whole number** is a number that we use for counting and numbering. We use these ten symbols, called **digits**, to represent whole numbers:

0, 1, 2, 3, 4, 5, 6, 7, 8, and 9

Using these ten digits, we can express any number, large or small. Numbers above the ten basic digits are formed by using combinations of the ten basic digits. The number "10," for example, is formed by using a combination of the first two digits.

THE PLACE VALUE SYSTEM

We can determine the value of each digit in a number by looking at its **place**, or its location, in the number. Thus, each digit in a number has a **place value**. To illustrate place value, let's look at the figure, which shows the number "5" in three different *places*.

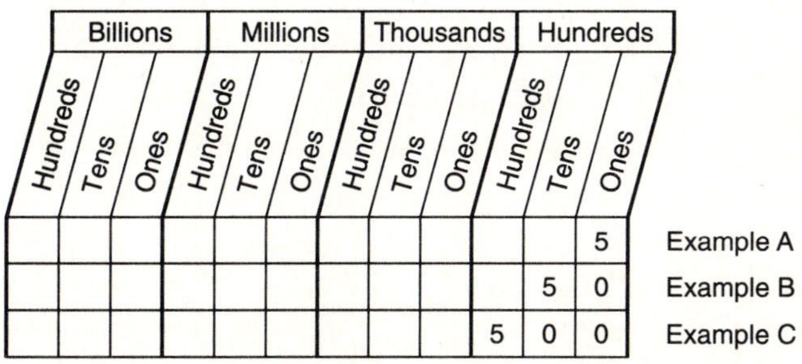

The first 5 (Example A) is in the *ones* place; therefore, it has a value of 5 ones. The second 5 (Example B) is in the *tens* place. So, it has a value of 5 tens and no ones, and is written as *50*. The third 5 (Example C) is in the *hundreds* place; so, it has a value of 5 hundreds, no tens, and no ones, and is written as *500*. Notice that each place has a value that is worth ten times the value of the place to its right.

The place value system lets us write numbers in a very compact way. For example, the number 256 is actually:

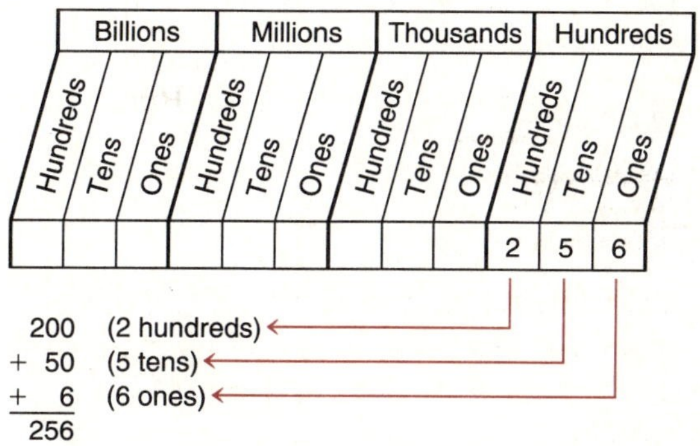

The number "256" is read as "two hundred fifty-six" and is the sum of its place values.

READING LARGER NUMBERS

Using the place value system makes it much easier to express and read large numbers. For example, let's look at the figure, which shows the number 1,216,566 with its place values labeled.

Billions			Millions			Thousands			Hundreds		
Hundreds	Tens	Ones	Hundreds	Tens	Ones	Hundreds	Tens	Ones	Hundreds	Tens	Ones
					1	2	1	6	5	6	6

We read this number as **one million, two hundred sixteen thousand, five hundred sixty-six.**

To make the number easier to read, commas are used to separate the digits into groups of three. Each group of three digits (except the ones group) is read by saying the three-digit number followed by the name of the group it belongs to. Let's look at some other examples.

EXAMPLE 1: Read the number 12,318,412 out loud.

Solution:

Billions			Millions			Thousands			Hundreds		
Hundreds	Tens	Ones	Hundreds	Tens	Ones	Hundreds	Tens	Ones	Hundreds	Tens	Ones
				1	2	3	1	8	4	1	2

twelve million, three hundred eighteen thousand, four hundred twelve

EXAMPLE 2: Read the number 10,615,258,016 out loud.

Solution:

Billions			Millions			Thousands			Hundreds		
Hundreds	Tens	Ones	Hundreds	Tens	Ones	Hundreds	Tens	Ones	Hundreds	Tens	Ones
	1	0	6	1	5	2	5	8	0	1	6

ten billion, six hundred fifteen million, two hundred fifty-eight thousand, sixteen

EXAMPLE 3: Read the number 210,344,865,008 out loud.

Solution:

Billions			Millions			Thousands			Hundreds		
Hundreds	Tens	Ones	Hundreds	Tens	Ones	Hundreds	Tens	Ones	Hundreds	Tens	Ones
2	1	0	3	4	4	8	6	5	0	0	8

two hundred ten billion, three hundred forty-four million, eight hundred sixty-five thousand, eight

In reading the number in the last example, notice that zeros are not read. This is because a zero means "no value." Thus, the number 500 means: no ones, no tens, and five hundreds. In the same way, the number 3,008 means: eight ones, no tens, no hundreds, and three thousands.

Also notice that we did not use the word "and" when reading any of the whole numbers in our examples. To take another example, the number "375" is read as "three hundred seventy-five," *not* as "three hundred and seventy-five." You only use "and" to show that a whole number also has a partial unit; we will talk about this in a later unit.

EXPRESSING WHOLE NUMBERS AS DOLLARS

To express a number in dollars, you write the dollar sign to the left of the number. Thus, "fifty dollars" is written as $50. To give another example, "four thousand, five hundred forty dollars" is written as $4,540.

Name ______________________ Date ______________________

UNIT 1 Exercises

Write the place value of the underlined digit.

1. 566
2. 287
3. 892
4. 1,975
5. 4,587
6. 23,498
7. 44,593
8. 125,986
9. 345,687
10. 15,677,890

ANSWERS

1. ______
2. ______
3. ______
4. ______
5. ______
6. ______
7. ______
8. ______
9. ______
10. ______

Write these numbers.

	Billions			Millions			Thousands			Hundreds		
	Hundreds	Tens	Ones	Hundreds	Tens	Ones	Hundreds	Tens	Ones	Hundreds	Tens	Ones
11.								2	2	8	1	5
12.					3	8	0	1	6	7	7	2
13.			5	6	7	9	4	3	8	2	9	9
14.	5	4	5	8	9	6	3	7	5	0	0	0
15.				1	0	0	3	1	8	0	1	7
16.							9	9	0	0	0	0

Write these numbers in words.

17. 76 ______________________

18. 235 ______________________

19. 2,398 ______________________

20. 12,398 ______________________

21. 123,498 ______________________________

22. 1,284,598 ______________________________

*23. 34,598,165 ______________________________

*24. 125,385,387 ______________________________

*25. 33,325,385,398 ______________________________

Write these numbers in digits.

ANSWERS

26. seventy-five

27. [illegible]ty-eight

28. two hundred ninety-nine

29. Eight hundred four

30. Fifteen thousand, three hundred three

31. One hundred two thousand, thirty-nine

32. Six million, four hundred two thousand

26. ____________
27. ____________
28. ____________
29. ____________
30. ____________
31. ____________
32. ____________

UNIT 2

Rounding Off Whole Numbers

Astronomers tell us that the sun is 93,000,000 miles from earth. Does this mean that the distance has been exactly measured and the sun is precisely 93,000,000 miles away? Or does it mean that, by using mathematical formulas, astronomers have calculated that the sun is around 93,000,000 miles away and this figure is rounded to even millions to make it easier to work with and to remember?

Actually, the sun is *around* 93,000,000 miles from earth, not *exactly* 93,000,000 miles. There are many times when numbers are rounded to make them easier to use. For example, the population of New York City is often said to be 8,000,000 people. At a given time, however, the actual population could be a little over 8,000,000 or a little under 8,000,000. What if it were, say, 8,124,512? This number would be hard to remember. So, it is simply rounded to an even 8,000,000.

To "round off" a number means to use an approximate value instead of the exact value. To round off a number, follow these steps:

Step 1: Locate the place that you want to round the number to.

Step 2: Look at the digit to the immediate *right* of the digit you want to round:

- If the digit to the right of the place being rounded is 5 (or more), add 1 to the place being rounded.
- If the digit to the right of the place being rounded is less than 5, do not change the digit in the place being rounded.

Step 3: Drop all digits after the place being rounded and replace them with zeros.

EXAMPLE 1: Round 48[illegible] to the nearest ten

Solution:

Step 1: 485 — Locate [illegible] tens place.

Step 2: 485 → 495 — Look at th[illegible] digit to the right of the [illegible]ns place. Since this d[illegible] is 5, the digit in the te[illegible] place, 8, is increased b[illegible]: 8 + 1 = 9.

Step 3: 495 → 490 — Change the dig[illegible]fter the tens place to zero.

Rounded answer = 490

EXAMPLE 2: Round 77,242 to the nea[illegible]t thousand

Solution:

Step 1: 77,241 — Locate the thousands place.

Step 2: 77,241 — The digit to the right of the thousands place is 2, which is less than 5; do not change the digit in the thousands place.

Step 3: 77,241 → 77,000 — Change all digits after the rounding place to zero.

Rounded answer = 77,000

It is important to remember that **when rounding, you only look at two digits: the place being rounded and the place to its right; do not look at any other digit.**

Now, before we start the exercises, let's look at a few more examples.

EXAMPLES:

788 rounded to the nearest ten = 790

7,455 rounded to the nearest hundred = 7,500

12,898 rounded to the nearest thousand = 13,000

255,790 rounded to the nearest ten-thousand = 260,000

397,800 rounded to the nearest hundred-thousand = 400,000

1,897,476 rounded to the nearest million = 2,000,000

Name ______________________ Date ______________________

UNIT 2 Exercises

Round these numbers to the nearest ten.

1. 75
2. 874
3. 999
4. 899
5. 3,456
6. 9,593
7. 12,356
8. 15,837
9. 25,925
10. 33,121

Round these numbers to the nearest hundred.

11. 678
12. 234
13. 1,565
14. 1,843
15. 2,877
16. 4,610
17. 12,875
18. 17,008
19. 21,081
20. 34,571

Round these numbers to the nearest thousand.

21. 1,898
22. 2,458
23. 7,916
24. 5,718
25. 9,469
26. 21,967
27. 45,854
28. 67,858
29. 91,546
30. 77,999

ANSWERS

1. ________
2. ________
3. ________
4. ________
5. ________
6. ________
7. ________
8. ________
9. ________
10. ________
11. ________
12. ________
13. ________
14. ________
15. ________
16. ________
17. ________
18. ________
19. ________
20. ________
21. ________
22. ________
23. ________
24. ________
25. ________
26. ________
27. ________
28. ________
29. ________
30. ________

Round these numbers to the nearest ten-thousand.

31. 34,598

32. 56,987

33. 47,698

34. 59,308

35. 59,287

36. 18,387

37. 48,698

38. 187,487

39. 287,376

40. 155,985

Round these numbers to the nearest hundred-thousand.

41. 128,765

42. 768,876

43. 345,986

44. 234,765

45. 998,475

46. 705,982

47. 3,409,876

48. 4,986,487

49. 7,856,387

50. 8,265,973

Round these numbers to the nearest million.

51. 3,598,587

52. 4,598,219

53. 8,498,986

54. 3,999,876

55. 7,987,376

56. 12,398,459

57. 24,998,765

58. 56,985,876

59. 55,267,985

60. 79,999,000

ANSWERS

31. ____________
32. ____________
33. ____________
34. ____________
35. ____________
36. ____________
37. ____________
38. ____________
39. ____________
40. ____________
41. ____________
42. ____________
43. ____________
44. ____________
45. ____________
46. ____________
47. ____________
48. ____________
49. ____________
50. ____________
51. ____________
52. ____________
53. ____________
54. ____________
55. ____________
56. ____________
57. ____________
58. ____________
59. ____________
60. ____________

Name ______________________ Date ______________________

UNIT 2 Applications

1. Algood Company had these sales figures for the year:

Territory	Sales
North	$2,988,964
South	3,499,437
East	3,288,388
West	4,988,876
East Central	2,388,984

Round each figure to the nearest thousand dollars.

2. Below are this month's sales for Lakeside Grocery. Round each amount to the nearest ten-thousand dollars.

Grocery: $78,988

Meat: $54,987

Produce: $32,876

Dairy: $18,769

Hardware: $12,344

3. Round the cost of these clothing items to the nearest ten dollars.

Leather coats: $99

Suede coats: $88

Belts: $12

Dress shoes: $65

Top coats: $95

ANSWERS

1. ______________
____
____
____
____

2. ______________
____
____
____
____

3. ______________
____
____
____
____

4. Zen Corporation had these profit figures for the last five years. Round each figure to the nearest million dollars.

19X1 . . . . . . $2,488,987

19X2 . . . . . . $6,487,184

19X3 . . . . . . $7,765,548

19X4 . . . . . . $7,687,175

19X5 . . . . . . $7,123,465

5. Round 12,467,886 to the nearest:

a. ten

b. hundred

c. thousand

d. ten-thousand

e. hundred-thousand

f. million

4. ____________

____

____

____

____

5. a. ____________

b. ____________

c. ____________

d. ____________

e. ____________

f. ____________

UNIT 3 Adding Whole Numbers

Addition is an arithmetic operation in which two or more numbers (called **addends**) are combined into a single larger number called the **sum** or the **total.** You show addition by writing the *plus* symbol ("+") next to the addends. Let's look at a simple addition problem to review these terms:

```
    9  ←——— addend
+   6  ←——— addend
   15  ←——— sum or total
```

To add larger whole numbers, you must align the numbers to be added (addends) into columns with ones digits over ones digits, tens digits over tens digits, hundreds digits over hundreds digits, and so on. Let's look at an example.

EXAMPLE 1: Find the sum of 24 and 18.
Solution:

```
tens digits ↓↓ ones digits
     24
   + 18
```

The "4" in the number 24 means 4 ones; the "8" in the number 18 means 8 ones. So, these digits must be aligned before you can start adding. The "2" in the number 24 means 2 tens; the "1" in the number 18 means 1 ten. So these digits must be aligned.

Now, start the addition by adding the digits in the ones place (4 ones + 8 ones = 12). Write down the "2" under the ones column and carry the "1" to the top of the tens colun[illegible] then add the digits in the tens column:

```
   1  ←——— carry
  24
+ 18
  42
  ↑↑— 4 + 8 = 12; write 2[illegible]rry 1
  └── 1 + 2 + 1 = 4
```

Since the sum of the tens c[illegible]mn, 4, is less than 10, there is no carry. Let [illegible]ok at another example.

EXAMPLE 2: Find the su[illegible]f 1,345 and 928
Solution:

```
  1 1  ←——— ca[illegible]
  1,345
+   928
  2,273
  ↑↑↑↑— 5 + 8 = 1[illegible]rite 3, carry 1
  ││└── 1 + 4 + 2[illegible]7; no carry
  │└─── 3 + 9 = [illegible]write 2, carry 1
  └──── 1 + 1 = 2
```

CHECKING FOR ACCURACY

The accuracy (or correctness) of any arithmetic operation should always be checked. An easy way to check the accuracy of an addition problem is simply to add backwards. That is, you add downward to get your sum, then you add upward to check your answer.

EXAMPLE 3: Add and check: 348 + 295 + 125

Solution:

Add downward to get a sum ↓

$$\begin{array}{r} \underline{768} \\ 348 \\ 295 \\ \underline{125} \\ 768 \end{array}$$

↑ Add upward to check your answer

Another way to check the accuracy in an addition problem is to switch the order of the addends in a column and re-add the column. You will get the same answer if your work is correct because the order of the addends does not change the sum.

Name ____________________ Date ____________________

UNIT 3 Exercises

Add these numbers.

1. 45 15	2. 45 23	3. 47 18	4. 86 27	5. 56 38	6. 98 19
7. 86 41	8. 35 15	9. 87 78	10. 75 27	11. 67 39	12. 76 35
13. 456 34	14. 287 9	15. 387 126	16. 186 167	17. 823 638	18. 487 187
19. 898 198	20. 509 387	21. 678 287	22. 386 492	23. 487 224	24. 896 892
25. 3,598 808	26. 9,076 988	27. 5,398 3,455	28. 8,439 4,986	29. 3,365 2,974	

ANSWERS

1. ______
2. ______
3. ______
4. ______
5. ______
6. ______
7. ______
8. ______
9. ______
10. ______
11. ______
12. ______
13. ______
14. ______
15. ______
16. ______
17. ______
18. ______
19. ______
20. ______
21. ______
22. ______
23. ______
24. ______
25. ______
26. ______
27. ______
28. ______
29. ______

30. 8,794 5,986	31. 5,388 4,694	32. 4,588 2,499	33. 5,299 3,268	34. 8,766 4,563
35. 12,518 3,765	36. 35,877 2,008	37. 89,764 12,786	38. 65,747 54,865	39. 32,897 18,649
40. 345,611 55,833	41. 347,365 134,587	42. 946,127 577,387		

Find the sum and check for accuracy in each of these problems.

43. 234 + 129 + 19 =

44. 172 + 12 + 8 =

45. 976 + 874 + 912 =

46. 162 + 478 + 211 =

47. 125 + 644 + 901 =

48. 617 + 911 =

49. 4,597 + 2,345 + 876 + 77 =

ANSWERS

30. ____________
31. ____________
32. ____________
33. ____________
34. ____________
35. ____________
36. ____________
37. ____________
38. ____________
39. ____________
40. ____________
41. ____________
42. ____________
43. ____________
44. ____________
45. ____________
46. ____________
47. ____________
48. ____________
49. ____________

Name ______________________ Date ______________________

50. 2,488 + 1,276 + 1,145 + 221 =

51. 8,455 + 4,356 + 1,450 + 865 =

52. 7,745 + 2,938 + 2,386 + 75 =

53. 5,588 + 3,455 + 3,599 + 255 =

54. 9,348 + 7,459 + 2,499 + 1,235 =

55. 8,677 + 3,587 + 2,699 + 2,349 =

56. 3,582 + 77 + 8 + 5 =

ANSWERS

50. ______

51. ______

52. ______

53. ______

54. ______

55. ______

56. ______

57. 3,485 + 275 + 211 + 11 + 9 =

58. 12,455 + 2,455 + 918 + 212 =

59. 125,678 + 109,489 + 4,566 =

60. 345,689 + 234,587 + 99,877 =

ANSWERS

57. ____________
58. ____________
59. ____________
60. ____________
61. ____________
62. ____________
63. ____________
64. ____________
65. ____________
66. ____________
67. ____________
68. ____________
69. ____________
70. ____________

Add these numbers and check for accuracy.

61.	62.	63.	64.
25,890	38,925	45,611	68,957
12,911	22,344	21,549	12,417
8,452	7,894	3,578	932
6,784	2,375	3,247	222
657	946	76	17

65.	66.	67.	68.
34,599	45,587	89,476	124,599
24,563	34,294	45,934	104,598
3,459	22,345	56,398	99,458
2,388	11,855	12,345	12,765

69.	70.
1,245,918	3,455,398
1,009,213	999,347

Name ______________________ Date ______________________

UNIT 3 Applications

In each of these application problems, look for words that give you some clue that you have to use addition. These are words like sum, total, combined, and altogether.

ANSWERS

1. ____________
2. ____________
3. ____________
4. ____________

1. Wayne Burke bought these books for his classes:

Biology	$20
Math	12
English	18
History	28

What was the total cost of his books?

2. Susan Kaye made these bank deposits last month: $125, $275, $321, and $508. What was the total amount she deposited during the month?

3. Pete Leo bought a house for $38,800 and made these repairs:

Paint	$ 400
Wallpaper	390
Carpet	1,950
Storm windows	1,250
Tile	850

What was the final cost of his house?

4. Spencer Variety Store made these amounts last week: Monday, $12,233; Tuesday, $9,455; Wednesday, $12,478; Thursday, $14,569; Friday, $17,450; Saturday, $23,455. What were the total sales for the week?

5. Susan Chun teaches business classes. This term she is teaching these subjects:

Subject	Number of Students
Business Math	35
Bookkeeping	25
Typing	40
Business English	15

What is the total number of students she is teaching?

ANSWERS

5. ________

6. ________

7. ________

8. ________

6. On July 1, Dave Rodriguez bought 125 acres of land. On August 17, Dave bought another 500 acres. And on September 15, he bought 340 more acres. How many acres did Dave purchase altogether?

7. The Simmons Company owns property that is worth $125,800; the Dyers Company owns property that is worth $250,000. What would be the total value of the properties if the two companies were formed into one company?

8. Sue Lawson earns $900 a month and lives with her parents. She is thinking about moving into an apartment. She figures she will have these expenses:

Rent	$300
Food	200
Utilities	100
Clothes	50
Car payment	225

Is Sue earning enough to move out?

Name ______________________________ Date ______________________________

9. Tim Lowell, owner of The Lunch Box Cafe, has 58 pounds of ground beef on Friday afternoon. If Tim bought another 178 pounds, how much ground beef would he have for the weekend?

ANSWERS

9. ____________

10. ____________

11. ____________

12. ____________

10. Chi Yi has these business math test scores for this term: 88, 89, 95, 94, 91, 87. What is Chi's total number of points?

11. Ben Stein sells books. Last week, he drove this many miles:

Monday	—	124 miles
Tuesday	—	198 miles
Wednesday	—	72 miles
Thursday	—	157 miles
Friday	—	205 miles

How many total miles did Ben drive during the week?

12. Itasca Company, which has a chain of grocery stores, showed these sales for the last three months:

western region	—	$345,800
northeastern region	—	$542,875
midwestern region	—	$234,600
southeastern region	—	$423,545

What total sales did the company have for these months?

Work Space

UNIT 4

Combinations That Total Ten

Even in our modern age of credit card size calculators and briefcase size computers, there is a great need for the ability to manually add numbers with speed and accuracy. Very often you may find that you need to add numbers quickly, and a calculator or computer may not be handy. And many situations in our fast-moving business world may call for you to quickly add numbers in your head.

You can gain speed and accuracy in addition more quickly if you learn to recognize combinations of **two** and **three** numbers that total 10. Here are some combinations of numbers that add up to 10:

	1	2	3	4	5
+	9	8	7	6	5
	10	10	10	10	10

	1	1	1	1	2	3	4
	1	2	3	4	3	3	4
+	8	7	6	5	5	4	2
	10	10	10	10	10	10	10

You should learn these combinations so well that when you see them you will think 10 right away without having to mentally add: Note how quickly we can add the following numbers using combinations that add up to 10:

5, 5 > 10	3, 7 > 10	8, 2 > 10
6, 4 > 10	6 — 6	6
7, 3 > 10	5, 5 > 10	3 — 3
	2 — 2	3 — 3
		4 — 10
30	28	26

Larger numbers can also be added more easily if you look for combinations that total 10.

EXAMPLE: Add: 853 + 657 + 486 + 324

Solution:

```
   2  2  ←———— carry
   8  5  3
   6  5  7
   4  8  6
   3  2  4
 ———————————
2  3  2  0
```

- 10 + 10 = 20; write 0 and carry 2
- 2 + 10 + 10 = 22; write 2 and carry 2
- 10 + 10 + 3 = 23; write 23

Work Space

Name ______________________ Date ______________________

UNIT 4 Exercises

Add these numbers. Mark combinations that total 10 before adding.

1.	2.	3.	4.	5.	6.
8	7	5	6	73	73
2	3	5	4	37	35
6	5	6	7	82	65
4	5	4	3	28	47

7.	8.	9.	10.	11.	12.
12	23	14	44	77	32
18	77	46	44	37	68
83	16	52	22	56	72
47	64	38	11	54	48

13.	14.	15.	16.	17.	18.
567	287	985	769	346	2,566
543	883	125	341	764	8,444
287	287	187	198	975	7,863
865	986	165	12	135	3,286

19.	20.	21.	22.
3,765	3,985	7,984	12,344
3,345	2,825	3,126	18,766
2,488	3,766	5,788	17,328
8,763	2,469	5,322	10,922
			8,188
			2,024

ANSWERS

1. ____________
2. ____________
3. ____________
4. ____________
5. ____________
6. ____________
7. ____________
8. ____________
9. ____________
10. ____________
11. ____________
12. ____________
13. ____________
14. ____________
15. ____________
16. ____________
17. ____________
18. ____________
19. ____________
20. ____________
21. ____________
22. ____________

23. 28,975	24. 44,652	25. 47,653
12,135	26,458	13,477
27,774	83,921	11,737
83,336	67,189	99,373
14,638	5,433	3,999
6,472	5,678	7,101

ANSWERS

23. ____________

24. ____________

25. ____________

Name ______________________ Date ______________________

UNIT 4 Applications

In each application problem, look for combinations that total 10.

ANSWERS

1. Southside Grocery made these sales of apples during the past week:

   Monday . . . . 48 lb
   Tuesday . . . . 62 lb
   Wednesday . . . . . 75 lb
   Thursday . . . . 35 lb
   Friday . . . . 88 lb
   Saturday . . . . 92 lb

   What was the total amount of apples sold during the week?

1. ____________

2. Finish this sales report for Kite Company.

Sales Report
Kite Company

Salesperson	Monday	Tuesday	Wednesday	Thursday	Friday
S. Myers	$ 882	$ 978	$ 565	$ 763	$ 762
T. Raja	228	137	545	647	748
G. Goode	327	456	376	765	376
H. Danzio	783	654	934	345	678
Total	$	$	$	$	$

3. Finish filling in this production schedule:

Product	Units Produced				
	Monday	**Tuesday**	**Wednesday**	**Thursday**	**Friday**
A	167	128	736	891	387
X	943	182	287	298	776
H	487	298	198	187	761
M	428	287	912	923	876
O	222	823	198	187	234
Totals					

UNIT

5

Horizontal Addition

The addition problems we have worked with so far have involved vertical addition (because we added up and down). Numbers to be added, however, are not always in vertical columns; sometimes they are written horizontally, as in Example 1.

Since adding numbers horizontally usually takes longer than adding numbers vertically, you need to practice horizontal addition. Let's look at Example 1 to see how the process works.

EXAMPLE 1: Add: 7 + 5 + 8 + 6 + 9 + 2 + 4
Solution:

7 + 5 + 8 + 6 + 9 + 2 + 4

12 20 26 35 37 = 41

You should look for combinations of numbers that total 10 when adding horizontally. This will help you to work faster and with less mistakes.

EXAMPLE 2: Add: 8 + 2 + 7 + 3 + 9 + 1
Solution:

8 + 2 + 7 + 3 + 9 + 1

10 17 20 29 = 30

When the addends have more than one digit, you add ones to ones, tens to tens, hundreds to hundreds, and so on. Remember to carry when the total of a row is 10 or more.

EXAMPLE 3: Add: 33 + 55 + 18 + 12
Solution: Add the ones digits first:

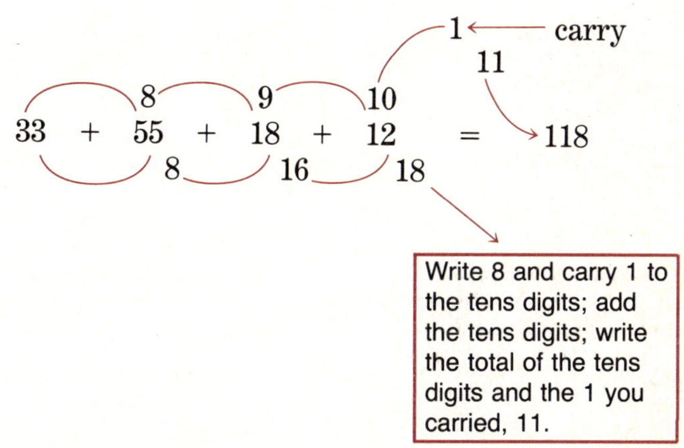

HORIZONTAL AND VERTICAL ADDITION

Sometimes you have to use both horizontal and vertical addition. Sales reports, for example, often call for the addition of horizontal rows and vertical columns. For example, let's look at the sales report of Clemmens Company for the week of June 5–9, 1991.

The total of the vertical columns should equal the total of the horizontal rows. This process, called crossfooting, is used to check for accuracy.

Clemmens Company
Sales Report
June 5–9, 1991

Salesperson	Monday	Tuesday	Wednesday	Thursday	Friday	Total
T. Lee	$ 250	$ 390	$1,250	$ 800	$ 760	$ 3,450
H. Taylor	1,200	700	800	450	540	3,690
D. Rodriguez	400	900	1,430	750	387	3,867
L. Sterling	1,050	560	830	780	420	3,640
Totals	$2,900	$2,550	$4,310	$2,780	$2,107	$14,647

Name ____________________ Date ____________________

UNIT 5 Exercises

Add these numbers horizontally.

1. 5 + 4 + 5 + 8 + 6 =

2. 4 + 4 + 5 + 5 + 9 =

3. 8 + 5 + 6 + 8 + 4 =

4. 6 + 5 + 3 + 9 + 7 =

5. 55 + 18 + 34 + 38 + 44 =

6. 39 + 65 + 68 + 26 + 38 =

7. 66 + 39 + 69 + 59 + 56 =

ANSWERS

1. ________
2. ________
3. ________
4. ________
5. ________
6. ________
7. ________

8. 125 + 457 + 139 + 225 =

9. 345 + 568 + 245 + 655 =

10. 455 + 235 + 540 + 545 =

11. 645 + 545 + 245 + 872 =

12. 2,450 + 1,288 + 850 =

13. 3,467 + 2,358 + 670 =

14. 4,565 + 2,455 + 3,187 =

ANSWERS

8. ____________

9. ____________

10. ____________

11. ____________

12. ____________

13. ____________

14. ____________

Name ______________________________ Date ______________________________

15. 8,990 + 5,675 + 2,111 =

ANSWERS

15. ________
16. ________
17. ________
18. ________
19. ________
20. ________
21. ________
22. ________

Add these problems horizontally and vertically. Check for accuracy by adding the totals of the vertical columns and horizontal rows.

16. 27 + 58 + 29 = ____
    18 + 28 + 21 = ____
    35 + 24 + 12 = ____
    45 + 23 + 20 = ____
    ___ + ___ + ___ = ____

17. 54 + 35 + 58 = ____
    87 + 32 + 25 = ____
    67 + 44 + 42 = ____
    78 + 58 + 66 = ____
    ___ + ___ + ___ = ____

18. 87 + 39 + 48 = ____
    25 + 46 + 26 = ____
    36 + 65 + 15 = ____
    17 + 8 + 6 = ____
    ___ + ___ + ___ = ____

19. 67 + 67 + 9 = ____
    43 + 23 + 12 = ____
    54 + 34 + 23 = ____
    48 + 32 + 23 = ____
    ___ + ___ + ___ = ____

20. 255 + 122 + 99 = ____
    365 + 276 + 232 = ____
    245 + 118 + 109 = ____
    856 + 455 + 318 = ____
    ___ + ___ + ___ = ____

21. 378 + 829 + 354 = ____
    244 + 388 + 329 = ____
    566 + 287 + 376 = ____
    455 + 376 + 176 = ____
    ___ + ___ + ___ = ____

22. 3,455 + 2,188 + 314 = ____
    2,387 + 3,498 + 2,897 = ____
    6,899 + 4,576 + 3,984 = ____
    8,976 + 2,589 + 3,006 = ____
    ___ + ___ + ___ = ____

Work Space

Name ______________________ Date ______________________

UNIT 5 Applications

Add the numbers horizontally in each of these problems.

ANSWERS

1. ________
2. ________
3. ________
4. ________

1. Jane Goodman works in a sporting goods store. On Tuesday, she made these sales:

   $55 $75 $88 $90 $8

   What are her total sales for the day?

2. The County Market showed these sales last week:

Monday	Tuesday	Wednesday	Thursday	Friday	Saturday
$7,565	$8,956	$12,514	$10,568	$9,997	$15,425

   Determine the total sales for the week.

3. Last week a secretary typed the following numbers of letters:

   15 8 9 12 6

   What was the total number of letters typed?

4. At a local power lifting contest, six power lifters entered the bench press category. Their individual presses were:

   290 lb 310 lb 285 lb 320 lb 301 lb 299 lb

   What was the total amount lifted by all six power lifters?

5. Tom Luchowski is a long distance runner. Last week he ran the following number of miles:

Mon., 12; Tue., 12; Wed., 15; Thurs., 10; Fri., 12; Sat., 14.

What was the total number of miles Bill ran?

ANSWERS

5. ____________

6. ____________

6. Express Delivery Company delivered the following number of packages last week:

Monday	Tuesday	Wednesday	Thursday	Friday	Saturday
1,200	1,950	3,455	1,966	2,840	2,955

What was the total number of packages delivered?

7. Complete this payroll report for Lakeview Company:

Lakeview Company
Payroll Report
For Week Ended June 25, 19XX

Hours Worked

Employee	Mon.	Tue.	Wed.	Thurs.	Fri.	Sat.	Total
L. Spears	8	8	10	10	7	5	______
D. McCollum	8	8	8	8	—	7	______
T. Yates	12	10	13	—	—	—	______
W. Wang	8	7	7	12	8	—	______
S. Day	11	10	8	8	—	4	______
K. Kaye	10	7	7	8	8	—	______
J. Robinson	8	8	8	8	8	—	______
Total	+	+	+	+	+	=	______

Name ______________________ Date ______________________

8. Complete this sales report for Henderson Products:

Henderson Products Weekly Sales Report May 2–6, 19XX						
Salesperson	**Monday**	**Tuesday**	**Wednesday**	**Thursday**	**Friday**	**Total**
Booker	$ 876	$1,200	$1,456	$ 978	$1,875	$______
Vouger	1,250	850	1,245	1,365	1,345	______
Simmons	1,436	1,243	987	963	1,422	______
O'Malley	997	1,233	1,577	1,678	1,003	______
Lang	1,375	1,876	1,986	998	1,245	______
Sterling	1,855	1,569	1,998	1,755	2,050	______
Total	+	+	+	+	=	______

9. Gateway Products Company sold the following numbers of items during the first six months of 19XX. Find the total number of items that were sold and fill out the sales report.

Gateway Products Company Sales Report For Period Ending June 30, 19XX							
Item	**January**	**February**	**March**	**April**	**May**	**June**	**Total**
A-9	550	986	376	588	612	618	______
B-12	890	854	789	987	980	897	______
B-11	998	887	912	876	670	698	______
C-1	786	764	876	981	576	886	______
D-14	312	413	234	543	254	256	______
D-18	121	213	122	139	298	276	______
Total	+	+	+	+	+	=	______

Work Space

UNIT 6 Subtracting Whole Numbers

Subtraction is an arithmetic operation in which you find the difference between two numbers. The number from which another is subtracted (or taken away) is called the **minuend,** and the number that is subtracted from the minuend is the **subtrahend.** The answer is called the **difference** or the **remainder.**

You show subtraction by writing the *minus* symbol ("−") in front of the subtrahend. Let's look at a simple problem to review these terms.

$$\begin{array}{rl} 8 \leftarrow & \text{minuend} \\ -\underline{3} \leftarrow & \text{subtrahend} \\ 5 \leftarrow & \text{difference} \end{array}$$

When setting up a subtraction problem (like with addition), be sure to line up the digits correctly: ones digits over ones digits, tens digits over tens digits, hundreds digits over hundreds digits, and so on. Let's look at two other examples.

EXAMPLE 1: Properly line up and subtract: 475 − 261

Solution:

$$\begin{array}{r} 475 \\ -\underline{261} \\ 214 \end{array}$$

5 − 1 = 4

7 − 6 = 1

4 − 2 = 2

EXAMPLE 2: Subtract: 8,756 −445

Solution:

$$\begin{array}{r} 8{,}756 \\ -\underline{\quad 445} \\ 8{,}311 \end{array}$$

6 − 5 = 1

5 − 4 = 1

7 − 4 = 3

8 − 0 = 8

Note that there is no digit in the thousands place of the subtrahend. So 8 minus zero equals 8.

CHECKING FOR ACCURACY IN A SUBTRACTION PROBLEM

You can check for accuracy in a subtraction problem by adding the difference to the subtrahend; the answer will equal the minuend. For example, we can check the accuracy of the answer in Example 2 (above) as follows:

$$\begin{array}{rl} 8{,}311 & \text{difference} \\ +\underline{\quad 445} & \text{subtrahend} \\ 8{,}756 & \text{minuend} \end{array}$$

Since the difference, 8,311, plus the subtrahend, 445, equals the minuend, 8,756, our arithmetic is correct.

Work Space

Name ______________________ Date ______________________

UNIT 6 Exercises

Find the difference in each of these problems.

1. 45 − 4
2. 68 − 7
3. 26 − 21
4. 65 − 42
5. 88 − 23
6. 95 − 75
7. 35 − 11
8. 25 − 14
9. 569 − 325
10. 432 − 120
11. 947 − 336
12. 222 − 201
13. 445 − 442
14. 886 − 675
15. 356 − 42
16. 456 − 5
17. 899 − 722
18. 875 − 632
19. 795 − 44
20. 698 − 472

Find the difference in each of these problems. Check by adding the difference to the subtrahend.

21. 468 − 345
22. 536 − 324
23. 9,837 − 526
24. 6,458 − 2,455
25. 6,594 − 5,482

ANSWERS

1. ________
2. ________
3. ________
4. ________
5. ________
6. ________
7. ________
8. ________
9. ________
10. ________
11. ________
12. ________
13. ________
14. ________
15. ________
16. ________
17. ________
18. ________
19. ________
20. ________
21. ________
22. ________
23. ________
24. ________
25. ________

26.	27.	28.	29.	30.
12,345	18,549	32,568	76,458	335,842
2,235	7,438	21,422	55,335	125,611

31.	32.	33.	34.	35.
896,995	766,455	45,986	699,314	20,211
875,674	256,311	4,222	44,204	201

Subtract each of these problems as shown.

36. 57 – 42   37. 68 – 14   38. 76 –16   39. 55 – 42

40. 225 – 15   41. 445 – 32   42. 665 – 432   43. 876 – 65

44. 2,566 – 435   45. 8,956 – 834   46. 7,992 – 5,751

47. 12,845 – 2,714   48. 77,218 – 25,115   49. 677,887 – 776

50. 869,875 – 535,224

ANSWERS

26. ____________
27. ____________
28. ____________
29. ____________
30. ____________
31. ____________
32. ____________
33. ____________
34. ____________
35. ____________
36. ____________
37. ____________
38. ____________
39. ____________
40. ____________
41. ____________
42. ____________
43. ____________
44. ____________
45. ____________
46. ____________
47. ____________
48. ____________
49. ____________
50. ____________

Name ______________________ Date ______________________

UNIT 6 Applications

1. On John's last visit to his doctor, he was told that he should weigh 175 pounds. If John now weighs 196 pounds, how much weight must he lose to get to his goal weight?

ANSWERS

1. ____________
2. ____________
3. ____________
4. ____________

2. Jenny Welsher earned $22,512 last year. This year she earned $27,855. How much more did she earn this year than last year?

3. The Ingrum family is spending $490 each month on groceries. If this is $40 more per month than they spent last year, how much a month were they spending a year ago?

4. After taxes, Bill Neeley earned $775 last month. He paid these bills:

Rent	$250
Power	130
Food	250
Car	75

How much did Bill have left after paying his bills?

5. On June 1, Susan Renike had $2,150 in her savings account. On June 15, she deposited another $300. But on June 25 she withdrew $250 for an unexpected bill. How much money did she have left after her withdrawal on June 25?

ANSWERS

5. ____________
6. ____________
7. ____________
8. ____________
9. ____________

6. William and Kathy Morris are planning a vacation trip to Florida. If they take the scenic route, they will have to drive 895 miles. But if they go by interstate, they will only have to drive 761 miles. How many miles will they save by taking the interstate?

7. In 1970 Leigh Taylor bought a house for $50,000. Because of inflation, the house is now worth $165,000. How much has the value of the house gone up?

8. Last year Macon Products Company reported total product sales of $345,233. This year the company reported total product sales of $466,345. How much did sales increase?

9. Phillip Cornett earned $598 before his employer took out any amounts. But his employer took out payroll deductions in the amount of $187. What was Phillip's "take-home" pay?

Name ______________________ Date ______________________

10. These tables show the selling price of certain items, and the cost of the same items, at McLean's Department Store. Find the amount of profit (selling price minus cost) on each item and the total profit.

(a)

Men's Clothing			
Item	**Selling Price**	**Cost**	**Profit**
Leather coats	$195	$125	$____
Sport coats	99	59	____
Wool ties	19	10	____
Silk ties	29	15	____
Dress shoes	35	21	____
Wool pants	75	44	____
Cotton pants	38	22	____
Belts	18	12	____
Totals	$____ –	$____ =	$____

(b)

Women's Clothing			
Item	**Selling Price**	**Cost**	**Profit**
Blouses	$ 37	$ 22	$____
Coats	199	107	____
Dresses	49	25	____
Pants	29	18	____
Hats	69	52	____
Leather purses	99	74	____
Skirts	38	24	____
Shoes	49	28	____
Totals	$____ –	$____ =	$____

Work Space

UNIT 7

Subtracting When Borrowing Is Needed

In Unit 6 we reviewed the terminology of a subtraction problem and we reviewed basic subtraction facts. We worked with subtraction problems such as:

$$\begin{array}{r} 986 \\ -\underline{321} \\ 665 \end{array}$$

Subtraction problems of this type usually give us little trouble because each digit in the minuend (remember, the minuend is the top number) is larger than the corresponding digit in the subtrahend. But when a digit in the subtrahend is larger than the corresponding digit in the minuend, you must borrow from the digits on the left. For example,

$$\begin{array}{r} 972 \\ -\underline{458} \\ ? \end{array}$$

You cannot subtract 8 from 2; so you must borrow from the "7" in the column to the left.

We can subtract by using the technique of borrowing:

$$\begin{array}{rrr} & 6 & \\ 9 & \not{7} & {}^{1}2 \\ -4 & 5 & 8 \\ \hline 5 & 1 & 4 \end{array}$$

Borrow 1 from the 7, making it 6 and adding 10 to the 2, making it 12. You add 10 to the 2 because you borrowed from the tens column: 12 − 8 = 4.

6 − 5 = 1

9 − 4 = 5

Let's look at two more examples.

EXAMPLE 1: Subtract and check: 634 − 395

Solution:

$$\begin{array}{rrr} 5 & {}^{1}2 & \\ \not{6} & \not{3} & {}^{1}4 \\ -3 & 9 & 5 \\ \hline 2 & 3 & 9 \end{array}$$

Borrow 1 from the 3, making it 2 and changing the 4 to 14.

14 − 5 = 9

Now borrow 1 from the 6, making it 5 and changing the 2 to 12: 12 − 9 = 3.

5 − 3 = 2

Check:

$$\begin{array}{rl} 395 & \text{subtrahend} \\ +\underline{239} & \text{difference} \\ 634 & \text{minuend} \end{array}$$

EXAMPLE 2: Subtract and check: 4,528 − 576

Solution:

$$\begin{array}{rcrrr} 3 & & {}^{1}4 & & \\ \not{4} & , & \not{5} & {}^{1}2 & 8 \\ & & 5 & 7 & 6 \\ \hline 3 & , & 9 & 5 & 2 \end{array}$$

8 − 6 = 2

Borrow 1 from the 5, making it 4 and changing the 2 to 12: 12 − 7 = 5.

Borrow 1 from the 4, making it 3 and changing the 4 to 14: 14 − 5 = 9.

3 − 0 = 3.

Check:

$$\begin{array}{rl} 3{,}952 & \text{subtrahend} \\ +\ \underline{576} & \text{difference} \\ 4{,}528 & \text{minuend} \end{array}$$

BORROWING FROM ZERO

In many subtraction problems the digit you need to borrow from will be a 0. When this happens, you must move to the left until you find a digit that is not 0 and borrow from that digit.

EXAMPLE 1: Subtract: 602 − 428.
Solution: First, set up the problem correctly:

$$\begin{array}{rccc} & 6 & 0 & 2 \\ - & \underline{4} & \underline{2} & \underline{8} \end{array}$$

Now, you need to borrow so that the 2 becomes 12. But you cannot borrow from the 0. So move over to the 6; borrow 1 from the 60: 60 − 1 = 59.

$$\begin{array}{rccc} & 5 & 9 & \\ & \not6 & \not0 & {}^{1}2 \\ - & \underline{4} & \underline{2} & \underline{8} \\ & 1 & 7 & 4 \end{array}$$

12 − 8 = 4

9 − 2 = 7

5 − 4 = 1

Check:

$$\begin{array}{rr} & 428 \\ + & \underline{174} \\ & 602 \end{array}$$

EXAMPLE 2: Subtract: 8,007 − 548

Set up the problem correctly:

$$\begin{array}{rccccc} & 8 & , & 0 & 0 & 7 \\ - & & & \underline{5} & \underline{4} & \underline{8} \end{array}$$

You must borrow so that the 7 becomes 17. Move to the left to the 8 and borrow 1 from the 800: 800 − 1 = 799:

$$\begin{array}{ccccc} 7 & & 9 & 9 & \\ \not8 & , & \not0 & \not0 & {}^{1}7 \\ \underline{} & & \underline{5} & \underline{4} & \underline{8} \\ 7 & , & 4 & 5 & 9 \end{array}$$

17 − 8 = 9

9 − 4 = 5

9 − 5 = 4

7 − 0 = 7

Check:

$$\begin{array}{rr} & 7{,}459 \\ + & \underline{548} \\ & 8{,}007 \end{array}$$

Name ________________________________ Date ________________________________

UNIT 7 Exercises

Find the difference for each of these problems.

1. 59	2. 69	3. 97	4. 38	5. 59	6. 83	7. 87
44	56	78	26	45	72	74

8. 59	9. 49	10. 95	11. 86	12. 75	13. 456	14. 678
28	37	38	62	33	45	56

15. 561	16. 323	17. 687	18. 5,697	19. 8,925
501	229	459	565	2,824

20. 15,865
12,845

Find the difference in each of these problems. Check by adding the subtrahend to the difference.

21. 342	22. 569	23. 5,680	24. 8,950	25. 8,235
19	278	762	2,875	4,922

ANSWERS

1. ________
2. ________
3. ________
4. ________
5. ________
6. ________
7. ________
8. ________
9. ________
10. ________
11. ________
12. ________
13. ________
14. ________
15. ________
16. ________
17. ________
18. ________
19. ________
20. ________
21. ________
22. ________
23. ________
24. ________
25. ________

26. 12,566 3,817	27. 14,876 12,935	28. 33,891 25,673	29. 145,225 34,874
30. 355,912 128,490	31. 548,298 374,985	32. 598,395 389,678	33. 384,937 298,543
34. 125,000 12,250	35. 345,985 234,876	36. 685,598 592,875	37. 2,345,984 978,654
38. 5,678,340 2,978,236	39. 12,340,986 4,568,985	40. 24,598,654 15,437,979	

Subtract and check each of these problems.

41. 72−49 42. 82−19 43. 57−28 44. 61−18 45. 553−78

46. 247 − 99 47. 554 − 28 48. 667 − 229 49. 931 − 286

50. 593 − 566 51. 576 − 399 52. 746 − 577

ANSWERS

26. ____________
27. ____________
28. ____________
29. ____________
30. ____________
31. ____________
32. ____________
33. ____________
34. ____________
35. ____________
36. ____________
37. ____________
38. ____________
39. ____________
40. ____________
41. ____________
42. ____________
43. ____________
44. ____________
45. ____________
46. ____________
47. ____________
48. ____________
49. ____________
50. ____________
51. ____________
52. ____________

Name ______________________ Date ______________________

53. 3,411 − 2,450 54. 5,957 − 2,516 55. 15,000 − 1,455

56. 25,205 − 22,462 57. 79,002 − 56,987 58. 125,890 − 12,769

59. 345,987 − 297,759 60. 500,000 − 126,988

ANSWERS

53. ____________
54. ____________
55. ____________
56. ____________
57. ____________
58. ____________
59. ____________
60. ____________

Work Space

Name ______________________ Date ______________________

UNIT 7 Applications

ANSWERS

1. ____________
2. ____________
3. ____________
4. ____________
5. ____________

1. A football stadium has 45,315 seats. At the last game, 33,488 tickets were sold. How many seats were empty?

2. Lakeside High School has 427 seniors and Eastdale High School has 389 seniors. How many more seniors are there at Lakeside High than Eastdale High?

3. The price of a new house is $87,500. If the house can be bought for $5,675 less, what will be the actual price of the house?

4. Kwon Lee earned $300 last week and her payroll deductions were: federal income taxes, $32; state income taxes, $9; social security taxes, $21. What was the actual amount of Kwon's check after all deductions were taken out?

5. In a recent mayor's race, one candidate got 33,458 votes while the other candidate got 24,597 votes. What was the difference in votes between the two candidates?

6. Mary Rose's electric bill went up from $99 in December to $127 in January. How much did her bill go up?

7. Bob Kaye was on a diet for the first two months of this year. At the start of his diet Bob weighed 227 pounds, and at the end of his diet Bob weighed 193 pounds. How much weight did he lose?

8. In January, Paul Rodriguez earned sales commissions of $5,675, and in February he earned sales commissions of $6,397. How much more did Paul earn in February?

9. At the end of the spring term, Central College raised its quarterly tuition from $137 to $155. How much did tuition go up?

10. Meri Robinson's total grocery bill was $75. But Meri had cash-back coupons for $2 and $5. What was Meri's bill after subtracting her cash-back coupons?

ANSWERS

6. __________

7. __________

8. __________

9. __________

10. __________

UNIT 8 Multiplying Whole Numbers

Multiplication is a shortcut to adding numbers. When you multiply numbers, you are actually adding groups of numbers. For example, multiplying 3 times 4, whose product is 12, is the same thing as adding 4 together 3 times (4 + 4 + 4 = 12).

Let's look at how a multiplication problem is set up and review its terms.

$$\begin{array}{rrl} & 23 & \leftarrow \text{multiplicand} \\ \times & \underline{3} & \leftarrow \text{multiplier} \\ & 69 & \leftarrow \text{product} \end{array}$$

Note that, as in addition and subtraction, multiplication problems are set up by correctly lining up digits: ones digits over ones digits, hundreds digits over hundreds digits, and so on. Also note that the number being multiplied is called the **multiplicand** and the number by which you multiply the multiplicand is called the **multiplier.** The answer in a multiplication problem is called the **product.**

Let's look at another example.

EXAMPLE: Multiply 345 × 4

Solution:

Step 1: Set up the problem correctly:

$$\begin{array}{rr} & 345 \\ \times & \underline{4} \end{array}$$

Step 2: Start multiplying from right to left: 4 × 5 = 20. Write 0 and carry 2 to the top of the next column:

$$\begin{array}{rrl} & 2 & \leftarrow \text{carry} \\ & 345 & \\ \times & \underline{4} & \\ & 0 & \end{array}$$

Step 3: Continue multiplying: 4 × 4 = 16, plus the 2 you carried equals 18. Write 8 and carry 1 to the top of the next column.

$$\begin{array}{rrl} & 12 & \leftarrow \text{carry} \\ & 345 & \\ \times & \underline{4} & \\ & 80 & \end{array}$$

Step 4: Finish doing your multiplication: 4 × 3 = 12, plus the 1 you carried equals 13. Write 13.

$$\begin{array}{rr} & 12 \\ & 345 \\ \times & \underline{4} \\ & 1{,}380 \end{array}$$

USING A MULTIPLICATION TABLE

The best way to improve your speed and accuracy when multiplying is to be sure that you know your multiplication tables. The following table can be used to practice your basic multiplication skills.

Basic Multiplication Facts

x	0	1	2	3	4	5	6	7	8	9
0	0	0	0	0	0	0	0	0	0	0
1		1	2	3	4	5	6	7	8	9
2			4	6	8	10	12	14	16	18
3				9	12	15	18	21	24	27
4					16	20	24	28	32	36
5						25	30	35	40	45
6							36	42	48	54
7								49	56	63
8									64	72
9										81

To illustrate how to use the table, let's say that you want to multiply 7 × 8. First, find the 7 *under* the column marked "x" at the extreme left. Then, find the 8 in the very top row to the right of the "x." Read across from the 7 and down from the 8 until you meet; you will see the number 56. So 7 × 8 = 56.

Notice that it is not necessary to show all products in the table. This is because 7 × 8 is the same as 8 × 7; the order of multiplication does not matter. Also notice that multiplying by 1 does not change the other number, and that multiplying by 0 always gives a product of 0.

Name ______________________ Date ______________________

UNIT 8 Exercises

Multiply each of these problems.

1. 45 × 2	2. 65 × 5	3. 69 × 8	4. 75 × 3	5. 26 × 6	6. 98 × 9	7. 54 × 2
8. 65 × 4	9. 76 × 7	10. 87 × 8	11. 58 × 7	12. 48 × 5	13. 39 × 9	14. 49 × 4
15. 49 × 6	16. 37 × 6	17. 36 × 8	18. 93 × 7	19. 97 × 6	20. 44 × 4	21. 48 × 8
22. 59 × 7	23. 96 × 5	24. 57 × 3	25. 56 × 4	26. 38 × 6	27. 96 × 8	28. 22 × 7

ANSWERS

1. ________
2. ________
3. ________
4. ________
5. ________
6. ________
7. ________
8. ________
9. ________
10. ________
11. ________
12. ________
13. ________
14. ________
15. ________
16. ________
17. ________
18. ________
19. ________
20. ________
21. ________
22. ________
23. ________
24. ________
25. ________
26. ________
27. ________
28. ________

29. 235	30. 455	31. 983	32. 485	33. 584	34. 298
5	9	6	8	7	9

35. 395	36. 586	37. 498	38. 684	39. 786	40. 899
2	8	4	9	8	3

41. 394	42. 597	43. 319	44. 477	45. 654	46. 3,566
6	7	3	7	8	8

47. 4,594	48. 5,938	49. 6,398	50. 7,976	51. 8,497
5	7	8	7	2

52. 3,498	53. 5,287	54. 5,688	55. 9,766	56. 12,538
9	4	7	9	4

57. 25,674	58. 55,655	59. 67,398	60. 34,877
6	8	5	9

ANSWERS

29. ____________
30. ____________
31. ____________
32. ____________
33. ____________
34. ____________
35. ____________
36. ____________
37. ____________
38. ____________
39. ____________
40. ____________
41. ____________
42. ____________
43. ____________
44. ____________
45. ____________
46. ____________
47. ____________
48. ____________
49. ____________
50. ____________
51. ____________
52. ____________
53. ____________
54. ____________
55. ____________
56. ____________
57. ____________
58. ____________
59. ____________
60. ____________

Name ______________________ Date ______________________

Using the table shown in this unit, multiply these numbers.

ANSWERS

61. $8 \times 9 =$ ______
62. $5 \times 8 =$ ______
63. $7 \times 7 =$ ______
64. $6 \times 9 =$ ______
65. $9 \times 9 =$ ______
66. $9 \times 0 =$ ______
67. $4 \times 8 =$ ______
68. $5 \times 7 =$ ______
69. $0 \times 9 =$ ______

61. ______
62. ______
63. ______
64. ______
65. ______
66. ______
67. ______
68. ______
69. ______

Work Space

Name ______________________ Date ______________________

UNIT 8 Applications

1. Bill Tai bought 8 lb of peanuts at $2 a pound. How much did he spend on peanuts?

2. A local baseball stadium has 7,552 seats. At $3 a ticket, how much money would be received on the night of a sellout?

3. Kathy Shelley just started a diet on which she feels she can lose 3 pounds each week. If the diet works, how much weight will she lose in 8 weeks?

4. If Tom Cecils makes $8 an hour, how much will he make in a week in which he worked 38 hours?

5. Mrs. Taggert's second grade class is going to give her a retirement party. If there are 30 students in the class, how much money will they have for the party if each student gives $2?

6. If a crate of oranges holds 756 oranges, how many oranges will 8 crates hold?

ANSWERS

1. ____________
2. ____________
3. ____________
4. ____________
5. ____________
6. ____________

7. Jane Ingalls is saving $25 each week. How much will she have in savings at the end of 8 weeks?

ANSWERS

7. ____________
8. ____________
9. ____________
10. ____________
11. ____________

8. If a pack of frozen waffles contains 6 waffles, how many waffles are there in 5 packs?

9. If movie tickets are $5 each, how much would it cost Bob Pierrot and 5 friends to go to a movie?

10. A local clothing store bought 355 leather belts at $7 each. What is the total cost of the belts?

11. If John Sansone puts $9 per week in his Christmas Club account for a whole year, how much will he have in the account by the end of the year?

12. Complete this daily earnings sheet for Madison Company.

Daily Earnings Sheet			
Employee	**Hourly Wage**	**Hours Worked**	**Total Earnings**
J. Myers	$ 8	8	$______
H. Booke	9	7	______
D. Swartz	12	8	______
K. Burke	10	6	______
B. Lasser	15	8	______
C. Bauer	11	7	______
T. Culman	14	8	______
B. Mayo	15	8	______
		Total	$______

UNIT 9

Multiplying Numbers with Two or More Digits in the Multiplier

When multiplying numbers you will often work with multipliers that have two or more digits. When this happens, you multiply the multiplicand (remember, the multiplicand is the top number) by *each* digit in the multiplier. Each separate multiplication gives you a **partial product.** You then add the partial products to get the final product.

EXAMPLE: Multiply: 645 × 35

Solution:

Step 1: Set up the problem correctly (remember, ones digits over ones digits, tens digits over tens digits, and so on):

$$\begin{array}{r} 645 \\ \times\ \underline{\ \ 35} \end{array}$$

Step 2: Multiply the multiplicand by the ones digit in the multiplier (5, in this problem):

$$\begin{array}{rl} 22 & \\ 645 & \\ \times\ \underline{\ \ 35} & \\ 3225 & \leftarrow \boxed{\text{partial product: } 5 \times 645 = 3{,}225} \end{array}$$

Step 3: Multiply the multiplicand by the tens digit in the multiplier (3, in this problem). You always put the first digit in a partial product right under the number you multiplied with. So write the first digit in this partial product right under the 3.

$$\begin{array}{rl} 11 & \\ 645 & \\ \times\ \underline{\ \ 35} & \\ 3225 & \\ 1935 & \leftarrow \boxed{\text{partial product: } 3 \times 645 = 1935} \end{array}$$

Step 4: Add the partial products to get the final product:

$$\begin{array}{rl} 645 & \\ \times\ \underline{\ \ 35} & \\ 3225 & \\ \underline{1935} & \\ 22575 & \leftarrow \text{product} \end{array}$$

Follow these same steps if you are multiplying larger numbers. Remember that each partial product is always written one place to the left of the partial product before it. Let's take another example.

EXAMPLE: Multiply: 3,455 × 324

Solution:

Step 1: Set up the problem correctly:

$$\begin{array}{r} 3{,}455 \\ \times\ \underline{\ \ 324} \end{array}$$

Step 2: Multiply the multiplicand by the ones digit in the multiplier (4, in this problem):

```
    1 22
   3,455
×    324
  13,820 ← 4 × 3,455 = 13,820
```

Step 3: Multiply the multiplicand by the tens digit in the multiplier (2, in this problem):

```
      11
   3,455
×    324
   13820
    6910 ← 2 × 3,455 = 6,910
```

Step 4: Multiply the multiplicand by the hundreds digit in the multiplier (3, in this problem):

```
    1 11
   3,455
×    324
   13820
    6910
  10365 ← 3 × 3,455 = 10,365
```

Step 5: Add the partial products to get the product:

```
     3,455
×      324
     13820
     6910
    10365
   1119420 = 1,119,420
```

CHECKING FOR ACCURACY IN A MULTIPLICATION PROBLEM

The best way to check the accuracy of a multiplication problem is to simply reverse your multiplication. As we learned in Unit 8, the order of multiplication does not matter. Let's look at an example.

MULTIPLY: 325 × 128 and check for accuracy by reverse multiplication.

Solution: **Accuracy check:**

```
     325                  128
×    128             ×    325
    2600                  640
    650                  256
   325                  384
   41600 ←——————————→   41600
```

Name ______________________ Date ______________________

UNIT 9 Exercises

Multiply each of these problems.

1. 46 × 25	2. 38 × 17	3. 59 × 49	4. 39 × 26	5. 98 × 84	6. 85 × 28
7. 87 × 27	8. 87 × 82	9. 28 × 22	10. 48 × 36	11. 87 × 59	12. 37 × 28
13. 97 × 82	14. 27 × 21	15. 89 × 43	16. 87 × 67	17. 41 × 39	18. 64 × 28
19. 387 × 82	20. 598 × 86	21. 387 × 45	22. 497 × 68	23. 298 × 38	

ANSWERS

1. ________
2. ________
3. ________
4. ________
5. ________
6. ________
7. ________
8. ________
9. ________
10. ________
11. ________
12. ________
13. ________
14. ________
15. ________
16. ________
17. ________
18. ________
19. ________
20. ________
21. ________
22. ________
23. ________

24. 398	25. 985	26. 487	27. 876	28. 984
86	32	49	83	61

29. 3,244	30. 4,877	31. 4,465	32. 8,945
378	327	445	765

33. 4,599	34. 5,388	35. 4,288	36. 9,451
241	218	653	298

37. 12,356	38. 54,388	39. 23,487	40. 36,871
566	294	492	593

ANSWERS

24. __________
25. __________
26. __________
27. __________
28. __________
29. __________
30. __________
31. __________
32. __________
33. __________
34. __________
35. __________
36. __________
37. __________
38. __________
39. __________
40. __________

Name ______________________ Date ______________________

Multiply these problems.

41. 345 × 48 42. 487 × 39 43. 387 × 42 44. 567 × 89

45. 487 × 265 46. 376 × 287 47. 438 × 236 48. 431 × 347

49. 478 × 211 50. 456 × 326 51. 459 × 322 52. 344 × 224

53. 918 × 118 54. 1,234 × 276 55. 2,238 × 287

ANSWERS

41. ______
42. ______
43. ______
44. ______
45. ______
46. ______
47. ______
48. ______
49. ______
50. ______
51. ______
52. ______
53. ______
54. ______
55. ______

56. $34{,}863 \times 485$ 57. $45{,}671 \times 287$ 58. $67{,}587 \times 287$

ANSWERS

56. ______________

57. ______________

58. ______________

59. ______________

60. ______________

59. $37{,}814 \times 567$ 60. $76{,}411 \times 388$

Name ______________________ Date ______________________

UNIT 9 Applications

1. If Bob Shiffman makes $12 an hour, what would he make in a week in which he worked 38 hours?

2. There is a type of candy that costs $16 a pound. How much would 15 pounds of candy cost?

3. Lolita can type 6 letters an hour. How many letters can she type in an 8-hour day?

4. If Ann's salary is $1,575 per month, what is her salary for a year?

5. A local sporting goods store bought 25 pairs of jogging shoes. If the price of each pair was $30, how much did the store pay to buy all of them?

6. Freemont High School's football stadium has 3,455 seats. If tickets sell for $3 each, how much would Freemont make on a night of a complete sellout?

ANSWERS

1. ____________
2. ____________
3. ____________
4. ____________
5. ____________
6. ____________

7. Louis bought 12 pounds of pecans at $10 a pound and 13 pounds of almonds at $4 a pound. What was the total amount that he spent?

ANSWERS

7. ____________

8. ____________

9. ____________

10. ____________

8. Lisa gets a commission of $25 on each membership she sells to the local health spa. If Lisa sold 16 memberships last week, what would be her total commission?

9. Three secretaries working together can stuff 325 envelopes an hour. How many envelopes can they stuff in 12 hours?

10. A local boy's club is having a cookout. They need 15 pounds of hamburger meat at $2 a pound, 25 packs of buns at $1 each, 8 packs of potato chips at $1 each, and 4 large bottles of cola at $2 each. What will be the total cost of the cookout?

11. Fill out this purchase order. Note: The symbol "@" means "at."

PURCHASE ORDER

To: Allied Office Supply Co.
Manchester, N.H.

No. 472
Date: 8/18/19XX

Quantity	Description	Amount
200 reams	20 lb bond paper @ $18 each	
18 dozen	yellow writing pads @ $12 dz	
12 dozen	#3 ballpoint pens @ $3 dz	
1	desktop calculator @ $125 each	
15	#125A computer ribbons @ $12 each	
18 boxes	correction fluid @ $5 box	
	Total	

Name ______________________ Date ______________________

12. Complete this invoice.

INVOICE

VICTORY AUTO PARTS COMPANY

To: Bill Stoops   Date: 11-6-19XX

Terms: Cash

Quantity	Description	Unit Price	Amount
50	#8 spark plugs	$ 2	$______
75	rearview mirrors	12	______
90	sideview mirrors	18	______
15	#120A blackwall tires	35	______
20	#128B whitewall tires	42	______
25	#45D floor mats	21	______
80	#2 deluxe seat cushions	9	______
		Total	$

Work Space

UNIT 10

Shortcut Multiplication

In Units 8 and 9 we reviewed multiplication of whole numbers. In this unit we are going to look at some shortcuts that can increase your speed and improve your accuracy when multiplying whole numbers.

MULTIPLYING WITH ZEROS IN THE MULTIPLIER

You should be careful when your multiplier has a zero (or zeros) in it because, as we discussed in Unit 9, any number multiplied by zero equals zero. Let's look at an example of multiplication when the multiplier has a zero.

EXAMPLE: Multiply 265 × 205
Solution:

```
    265
×   205
   1325 ←—— 5 × 265 = 1,325
   000  ←—— 0 × 265 = 000
  530   ←—— 2 × 265 = 530
 54,325
```

While this method works well, it often causes errors because it is easy to forget that 0 times any number is 0. We can make the multiplication easier by not multiplying by the 0. Instead, just bring down the zero into the product as a **placeholder.** Let's look at a couple of examples to see how this works.

EXAMPLE 1: Multiply: 265 × 205
Solution:

Step 1: Set up the problem correctly and multiply the multiplicand by the ones digits in the multiplier:

```
   32
  265
  205
1,325 ←—— 5 × 265 = 1,325
```

Step 2: Bring down the zero into the product as a placeholder:

```
  265
  205
1,325
    0
```

Step 3: Multiply the multiplicand by the hundreds digit in the multiplier:

```
  11
 265
 205
1325
5300 ←—— 2 × 265 = 530
```

Step 4: Add the partial products:

```
   265
   205
  1325
 5300
54,325
```

EXAMPLE 2: Multiply 5,225 × 2,004

Solution:

Step 1: Multiply the multiplicand by the ones digit in the multiplier:

```
   12
 5225
 2004
------
20,900 <——— 4 × 5,225 = 20,900
```

Step 2: Bring down the zeros into the product as placeholders:

```
 5,225
 2,004
------
 20900
   00
```

Step 3: Multiply the multiplicand by the thousands digit in the multiplier:

```
      1
  5,225
  2,004
--------
  20900
1045000 <——— 2 × 5,225 = 10450
```

Step 4: Add the partial products:

```
     5,225
     2,004
----------
     20900
   1045000
----------
10,470,900
```

MULTIPLYING BY A POWER OF TEN

When your multiplier is 10 or a power of 10 (for example, 100, 1,000, 10,000) and your multiplicand is a whole number, just count the zeros in the multiplier and add on the same number of zeros to the multiplicand.

EXAMPLE: Multiply: 55 × 10

Solution: 55 × 10 = 550

(10: one zero in the multiplier) (550: add on one zero to the multiplicand)

Let's look at a few more examples to see how well this works with larger numbers.

EXAMPLES: Multiply the following:

a. 657 × 1,000
b. 1,837 × 10,000
c. 12,456 × 100
d. 55,685 × 100,000

Solution:

a. 657 × 1,000 = 657,000 (add on three zeros)
b. 1,837 × 10,000 = 18,370,000 (add on four zeros)
c. 12,456 × 100 = 1,245,600 (add on two zeros)
d. 55,686 × 100,000 = 5,568,600,000 (add on five zeros)

MULTIPLYING WHEN THE MULTIPLIER ENDS WITH ZEROS

When the multiplier ends with a zero (or zeros), you can avoid multiplying by the zeros if you follow the steps in the next example.

EXAMPLE: Multiply: 515 × 200

Solution:

Step 1: Set the zeros to the right of the multiplicand:

```
515
  200
-----
```

Step 2: Bring zeros into the product:

```
515
  200
-----
   00
```

Step 3: Multiply by the nonzero whole numbers:

```
  1
515
  200
-------
103,000   <——— 2 × 515 = 1,030
```

Name ____________________ Date ____________________

UNIT 10 Exercises

Multiply each of these problems.

1. 274 × 50	2. 498 × 201	3. 137 × 40	4. 387 × 305	5. 476 × 404	6. 384 × 202
7. 298 × 120	8. 286 × 202	9. 985 × 501	10. 487 × 90	11. 2,455 × 405	12. 3,487 × 2,001
13. 1,265 × 900	14. 2,365 × 1,006	15. 2,379 × 2,031			

ANSWERS

1. ____________
2. ____________
3. ____________
4. ____________
5. ____________
6. ____________
7. ____________
8. ____________
9. ____________
10. ____________
11. ____________
12. ____________
13. ____________
14. ____________
15. ____________

Mentally multiply each of the following.

16. 561 × 10 =

17. 345 × 100 =

18. 481 × 1,000 =

19. 2,488 × 100 =

20. 459 × 10,000 =

21. 3,498 × 100,000 =

22. 12,456 × 100 =

23. 24,588 × 1,000 =

24. 51,344 × 10,000 =

25. 125,675 × 100 =

Multiply each of these problems.

26. 49 × 40 =

27. 345 × 50 =

28. 427 × 200 =

ANSWERS

16. ____________
17. ____________
18. ____________
19. ____________
20. ____________
21. ____________
22. ____________
23. ____________
24. ____________
25. ____________
26. ____________
27. ____________
28. ____________

Name ______________________ Date ______________________

29. $83 \times 6{,}000 =$

ANSWERS

29. ____________
30. ____________
31. ____________
32. ____________
33. ____________
34. ____________
35. ____________

30. $937 \times 500 =$

31. $1{,}564 \times 200 =$

32. $2{,}455 \times 700 =$

33. $18{,}255 \times 400 =$

34. $25{,}692 \times 900 =$

35. $555 \times 7{,}000 =$

ANSWERS

36. ____________
37. ____________
38. ____________
39. ____________
40. ____________

36. $3{,}522 \times 8{,}000 =$

37. $456 \times 250 =$

38. $4{,}598 \times 2{,}500 =$

39. $1{,}008 \times 500 =$

40. $2{,}500 \times 2{,}500 =$

Name ______________________ Date ______________________

► UNIT 10 Applications

Using the shortcuts you studied in this unit, solve each of these problems.

ANSWERS

1. ____________
2. ____________
3. ____________
4. ____________
5. ____________

1. Robert Larr bought 25 video games for $30 each. How much did it cost to buy all of these games?

2. There are 500 sheets in a ream of typing paper. How many sheets are in 50 reams?

3. During a special sale, a local clothing store sold 300 men's suits at $225 each. What was the total amount of the sale?

4. If envelopes are packaged 1,000 to a box, how many envelopes are in 50 boxes?

5. If one bottle of shampoo costs $2, how much would 200 bottles of the same shampoo cost?

6. A store owner bought 15 items for $100 each. What is the total cost of the purchase?

7. If Bill makes $1,200 a month, what does he make in a year?

8. Hanna Borkan makes $10 an hour. What will she make in a week in which she worked 40 hours?

9. Lakeside Fitness Company makes exercise equipment. Last month, Lakeside made 108 heavy-duty workout stations. If it cost $225 to make one station, what did it cost to make all 108 stations?

10. During a clearance sale, Duke Electronics sold 2,455 video recorders for $1,005 each. What was the total amount received for all sales?

11. Wickerstaff Company has 100 workers who each earn $10 an hour. What was the total factory payroll last week if each person worked 40 hours?

ANSWERS

6. ____________
7. ____________
8. ____________
9. ____________
10. ____________
11. ____________

Name ______________________ Date ______________________

12. Fill out this sales report for the Rosenberg Company.

Rosenberg Company—Sales Report			
Product	**Number of Units Sold**	**Unit Selling Price**	**Total Sale**
A	200	$ 18	$______
B	300	12	______
C	500	8	______
D	800	5	______
E	600	15	______
F	350	10	______
G	980	100	______
H	532	200	______
I	1,255	100	______
J	1,855	120	______
Totals	______		$______

Work Space

UNIT 11 Dividing Whole Numbers

If a pie is cut into six equal pieces to serve six people, we say that the pie has been *divided* six ways. In the same way, if a high school debate team that has 10 members is given $200 for expenses for a road trip, the $200 must be divided by 10. **Division** is the process of finding out how many times one number (the **divisor**) is contained in another number (the **dividend**). The answer in a division problem is called the **quotient.** Division is the opposite of multiplication. Let's look at a division problem and review its terms.

$$\text{divisor} \rightarrow 5\overline{)30} \leftarrow \text{dividend}, \quad 6 \leftarrow \text{quotient}$$

We know that 30 divided by 5 is 6 because 5 × 6 = 30. Remember, division is the opposite of multiplication.

In this example, we used the division box $\overline{)\quad}$ to indicate division. Another common way to indicate division is with the symbol "÷" which means "divided by." So, 30 ÷ 5 means the same thing as $5\overline{)30}$.

To find the answer to some division problems, you have to use long division. Let's take an example.

EXAMPLE: Divide 684 by 9

Solution:

Step 1: Set the problem up in a division box:

$$9\overline{)684}$$

Step 2: Look at the divisor (9) and the first digit in the dividend (6). How many 9's are there in 6? None. So move over to the 8 and ask how many 9's are in 68? There are seven 9's in 68, so you place the 7 directly above the 8.

$$\begin{array}{r} 7 \\ 9\overline{)684} \end{array} \qquad 68 \div 9 = 7$$

Step 3: Multiply 7 × 9 = 63. Write this product under the 68 in the dividend:

$$7 \times 9 = 63 \longrightarrow \begin{array}{r} 7 \\ 9\overline{)684} \\ \underline{63} \end{array}$$

Step 4: Subtract 63 from 68 to get a difference of 5 (68 − 63 = 5). Bring down the 4 from the dividend and write it next to the 5.

$$\begin{array}{r} 7 \\ 9\overline{)684} \\ \underline{63} \\ 54 \end{array} \leftarrow \text{new dividend}$$

Step 5: Now, you have a new dividend, 54. Divide the divisor, 9, into the new dividend. The result is 6 (54 ÷ 9 = 6). Write 6 in the quotient and multiply the 6 by the divisor.

Write the result (6 × 9 = 54) under the previous dividend of 54.

```
     76
 9 ⟌684
    63
    ---
    54
    54 <—— 6 × 9 = 54
    ---
     0
```

The last subtraction results in a difference of 0. So, 9 is contained in 684 exactly 76 times. When this happens, we say that the problem "comes out even." If a division problem does not come out even, there will be a "remainder" after you make your last subtraction. We will work with remainders in Unit 12.

CHECKING FOR ACCURACY IN DIVISION PROBLEMS

To check for accuracy in a division problem, you multiply the quotient by the divisor. The result, if your math is correct, will equal the dividend.

EXAMPLE: Divide 504 by 7 and check for accuracy.

Solution: **Accuracy check:**

```
     72
 7 ⟌504            72 <—— quotient
    49          ×   7 <—— divisor
    ---           ---
    14            504
    14
    ---
     0
```

Name ____________________ Date ____________________

UNIT 11 Exercises

Divide each of these problems.

1. $3\overline{)36}$ 2. $8\overline{)48}$ 3. $4\overline{)844}$ 4. $6\overline{)366}$ 5. $4\overline{)444}$

6. $5\overline{)2{,}555}$ 7. $7\overline{)6{,}377}$ 8. $8\overline{)6{,}496}$ 9. $6\overline{)4{,}284}$

10. $9\overline{)3{,}699}$ 11. $8\overline{)40{,}976}$ 12. $7\overline{)49{,}420}$ 13. $15\overline{)90}$

14. $16\overline{)112}$ 15. $68\overline{)340}$ 16. $25\overline{)200}$ 17. $45\overline{)360}$

ANSWERS

1. ________
2. ________
3. ________
4. ________
5. ________
6. ________
7. ________
8. ________
9. ________
10. ________
11. ________
12. ________
13. ________
14. ________
15. ________
16. ________
17. ________

18. $25\overline{)3{,}050}$ 19. $30\overline{)6{,}330}$ 20. $24\overline{)8{,}424}$ 21. $8\overline{)72{,}984}$

22. $214\overline{)642}$ 23. $147\overline{)735}$ 24. $225\overline{)1{,}800}$ 25. $301\overline{)1{,}806}$

26. $35\overline{)315}$ 27. $450\overline{)2{,}700}$ 28. $290\overline{)2{,}030}$ 29. $211\overline{)1{,}055}$

30. $30\overline{)2{,}850}$ 31. $75\overline{)24{,}375}$ 32. $12\overline{)6{,}264}$ 33. $725\overline{)6{,}525}$

ANSWERS

18. ____________
19. ____________
20. ____________
21. ____________
22. ____________
23. ____________
24. ____________
25. ____________
26. ____________
27. ____________
28. ____________
29. ____________
30. ____________
31. ____________
32. ____________
33. ____________

Name ____________________ Date ____________________

34. $650\overline{)11{,}700}$ 35. $123\overline{)15{,}375}$ 36. $636\overline{)28{,}620}$

ANSWERS

34. ________
35. ________
36. ________
37. ________
38. ________
39. ________
40. ________
41. ________
42. ________
43. ________
44. ________
45. ________

37. 18,396 ÷ 438 38. 17,136 ÷ 238 39. 17,435 ÷ 317

40. \$16,920 ÷ 235 41. \$28,305 ÷ 629 42. \$13,440 ÷ 384

43. \$27,170 ÷ 715 44. \$23,625 ÷ 315 45. \$75,525 ÷ 795

Work Space

Name ______________________ Date ______________________

▶ UNIT 11 Applications

1. Lori Kaye bought a new typewriter on credit for $1,236. If Lori made 12 equal payments, what was the amount of each payment?

ANSWERS

1. ____________
2. ____________
3. ____________
4. ____________

2. A local store bought 72 tables for $3,960. What is the cost of each table?

3. If 18 ounces of candy are divided among three children, how much would each child get?

4. A local television station is paying $13,000 for the rights to broadcast a semipro baseball game. If there are 20 players on each team (40 players altogether), how much would each player be paid?

5. Dave Hroncich makes $18,300 a year. What is his salary each month?

ANSWERS

5. ____________

6. ____________

7. ____________

8. ____________

6. A traveler drove 1,848 miles in 3 days. If he drove the same amount of distance each day, how many miles did he drive each day?

7. Lucy Algarin can run a mile in 6 minutes. At this pace, how many miles can she run in 18 minutes?

8. A new motorcycle costing $5,000 can be bought with a $500 down payment and the balance being paid in 36 monthly payments. What is the amount of each payment?

Name ______________________ Date ______________________

9. McGram Company pays $27,000 a month for health insurance on its 150 employees. What is the cost for each employee per month?

ANSWERS

9. ____________

10. ____________

11. ____________

12. ____________

10. What is the cost of 1 foot of copper tubing if 12 feet cost $24?

11. Lori Copeland earns $240 for a 40-hour work week. What does she make per hour?

12. Lakeside School bought 100 footballs for $500 total. What is the cost of one football?

13. The job title and yearly salaries for certain jobs in the Goozner Company are shown in the following form. Complete the form.

Job	Yearly Salary	Monthly* Salary	Weekly** Salary
Computer operator	$19,500	$______	$______
Computer programmer	39,000	______	______
Accountant	23,400	______	______
Secretary	11,700	______	______
Salesperson	35,100	______	______
Sales manager	46,800	______	______

*Divide yearly salary by 12
**Divide yearly salary by 52

UNIT 12 Division with a Remainder and Finding Averages

In Unit 11, we reviewed basic division facts and we reviewed the terminology of a division problem. We worked with division problems that "came out even." That is, the divisor divided evenly into the dividend without a final remainder.

In division problems, there will often be a remainder after you do the final subtraction. We call this **division with a remainder.**

EXAMPLE: Divide 658 by 8
Solution:

$$
\begin{array}{r}
82 \\
8\,\overline{)658} \\
\underline{64} \\
18 \\
\underline{16} \\
2
\end{array}
$$

remainder—because there is nothing left in the dividend to bring down

In this problem, we say that 658 divided by 8 equals 82 with "2 left over." So, the final answer can be stated as 82 R2 (which you read "82 with a remainder of 2"). Another way to show the remainder as part of the answer is to write it as a "fraction." To show a remainder as a fraction, draw a line under the remainder and write the divisor under the line. Then, write the fraction next to the quotient. As an example, let's look again at the above problem.

$$
\begin{array}{r}
82\frac{2}{8} \\
8\,\overline{)658} \\
\underline{64} \\
18 \\
\underline{16} \\
\frac{2}{8}
\end{array}
$$

fraction—another way to show division; $\frac{2}{8}$ also means $2 \div 8$ or $8\,\overline{)2}$

Now, let's look at a few more examples.

EXAMPLE 1: Divide 1,234 by 9 and express the remainder as a fraction.
Solution:

$$
\begin{array}{r}
137\frac{1}{9} \\
9\,\overline{)1{,}234} \\
\underline{9} \\
33 \\
\underline{27} \\
64 \\
\underline{63} \\
\text{remainder} \rightarrow \frac{1}{9}
\end{array}
$$

EXAMPLE 2: Divide 3,562 by 315 and express the remainder as a fraction.

Solution:

```
          11 97/315  <--+
    315 | 3562          |
          315           |
          ---           |
           412          |
           315          |
           ---          |
remainder --> 97 -------+
              ---
              315
```

CHECKING FOR ACCURACY WHEN THERE IS A REMAINDER

In the last unit, you learned that the accuracy of a division problem can be checked by multiplying the quotient by the divisor (this will yield the dividend). If the problem has a remainder, add the remainder to the product of the quotient and the divisor.

EXAMPLE: Divide 3,455 by 8 and check for accuracy.

Solution:

```
   431
8 | 3455
    32
    --
     25
     24
     --
      15
       8
      --
       7 <-- remainder
```

Accuracy check:

```
   431    quotient
×    8    divisor
  ----
  3448
+    7    remainder
  ----
  3455
```

FINDING AVERAGES

Many times in business, and in our personal lives, we may need to calculate an **average,** that is, a number that is representative of a group of numbers. For example, your final grade in a course is usually based on an "average" of all your test scores. Businesses often calculate such things as average daily sales, average monthly sales per salesperson, average number of miles traveled per trip, average daily balance of customers' accounts, and so on.

To find an average, follow these steps:

Step 1: Add together all of the numbers in a group.

Step 2: Divide this total by the number of numbers in the group.

EXAMPLE 1: John Brandwein had these scores in a business math class: 90, 95, 80, 94, and 91. Find his average score.

Solution:

Step 1: Find the total of all test scores:

$$90 + 95 + 80 + 94 + 91 = 450$$

Step 2: Divide the total of the test scores by the number of tests:

$$5\overline{)450} = 90$$, average score on 5 math tests

EXAMPLE 2: On Friday, six secretaries at Espon Supply Company typed 54 letters. What was the average number of letters typed by each secretary?

Solution: Divide 54 (the total number of letters typed) by 6 (the number of typists):

$$6\overline{)54} = 9$$, average number of letters typed

Name ______________________________ Date ______________________________

UNIT 12 Exercises

Divide each of these problems. Show any remainder as a fraction.

1. $4\overline{)37}$ 2. $9\overline{)95}$ 3. $5\overline{)27}$ 4. $3\overline{)19}$ 5. $7\overline{)45}$

6. $6\overline{)43}$ 7. $7\overline{)89}$ 8. $8\overline{)51}$ 9. $9\overline{)60}$ 10. $2\overline{)39}$

11. $5\overline{)49}$ 12. $7\overline{)71}$ 13. $3\overline{)95}$ 14. $9\overline{)88}$ 15. $6\overline{)57}$

16. $9\overline{)683}$ 17. $8\overline{)476}$ 18. $6\overline{)471}$ 19. $7\overline{)333}$

20. $4\overline{)933}$ 21. $12\overline{)447}$ 22. $15\overline{)221}$ 23. $18\overline{)657}$

ANSWERS

1. __________
2. __________
3. __________
4. __________
5. __________
6. __________
7. __________
8. __________
9. __________
10. __________
11. __________
12. __________
13. __________
14. __________
15. __________
16. __________
17. __________
18. __________
19. __________
20. __________
21. __________
22. __________
23. __________

24. $41\overline{)552}$ 25. $28\overline{)533}$ 26. $17\overline{)426}$ 27. $13\overline{)457}$

28. $21\overline{)675}$ 29. $19\overline{)477}$ 30. $41\overline{)618}$ 31. $33\overline{)595}$

32. $23\overline{)397}$ 33. $33\overline{)2{,}905}$ 34. $15\overline{)1{,}128}$ 35. $16\overline{)1{,}429}$

36. $25\overline{)2{,}378}$ 37. $27\overline{)1{,}947}$ 38. $21\overline{)1{,}871}$ 39. $19\overline{)1{,}655}$

ANSWERS

24. ________
25. ________
26. ________
27. ________
28. ________
29. ________
30. ________
31. ________
32. ________
33. ________
34. ________
35. ________
36. ________
37. ________
38. ________
39. ________

Name ______________________ Date ______________________

40. $17\overline{)1,618}$ 41. $385\overline{)21,177}$ 42. $218\overline{)11,337}$

ANSWERS

40. ________
41. ________
42. ________
43. ________
44. ________
45. ________
46. ________
47. ________
48. ________
49. ________
50. ________
51. ________

43. $715\overline{)34,322}$ 44. $614\overline{)52,295}$ 45. $313\overline{)29,737}$

46. $417\overline{)24,187}$ 47. $31,895 \div 612$ 48. $27,985 \div 318$

49. $12,569 \div 245$ 50. $91,179 \div 918$ 51. $123,125 \div 655$

52. 144,516 ÷ 612

ANSWERS

52. ____________
53. ____________
54. ____________
55. ____________
56. ____________
57. ____________
58. ____________

Find the average of each of these groups of numbers.

53. 45, 49, 38, 56

54. 92, 88, 78, 92, 80

55. 28, 48, 59, 69, 84

56. 125, 695, 485, 285, 598

57. $2,500, $5,600, $3,600

58. $125, $275, $95, $240

Name ______________________________ Date ______________________________

UNIT 12 Applications

ANSWERS

1. ____________
2. ____________
3. ____________
4. ____________
5. ____________

1. If a 12-case box of canned fruit weighs 46 pounds, what does each can weigh?

2. If 5 kids have even shares of an 18-ounce pizza, how much would each share weigh?

3. It takes 2 yards of fabric to make a certain kind of T-shirt. a. How many T-shirts can be made with 99 yards of fabric? b. How much fabric is left over?

4. A 6-member jogging team carried 7 quarts of juice with them on a cross-country run. At 32 ounces per quart, how many ounces of juice were there for each runner?

5. Bob Ziglar lost 37 pounds in 6 months. What was Bob's average weight loss for each month?

6. Ann Todd makes $325 for a 40-hour work week. What does she make an hour?

ANSWERS

6. ____________
7. ____________
8. ____________
9. ____________

7. Vicki and Mike Matli paid $7,350 for 600 square yards of carpet. What is the cost per square yard?

8. Ted Davis read that playing racquetball burns 400 calories an hour. How many hours would Ted need to play to burn 1,800 calories?

9. Five members of the Midtown Boy's Club held a bake sale to raise funds. The sale was a success but there were 16 pieces of pie left over. So, the boys divided up the leftover pies evenly. How much was each boy's share?

Name ________________________ Date ________________________

10. The Baker family is planning to drive 2,321 miles on their vacation. How many miles will they need to drive each day if they plan to make the trip in 3 days?

ANSWERS

10. ____________

11. ____________

12. ____________

11. Louis Lobo is a salesperson in a department store. Last week, Louis made these sales: Monday, $300; Tuesday, $320; Wednesday, $290; Thursday, $325; Friday, $250. What was the average of Louis' daily sales for the week?

12. The 1,200 workers at Harkon Industries built 72,000 units on Friday. How many units did each worker average?

Work Space

UNIT 13 Dividing When Zeros Are Needed in the Quotient

We have stressed that, in our system of place values, a zero contained in a number means no value in that place. For example, in the number 605, the 0 means "no value in the tens place." But even though the 0 means there are no tens, it is still needed as a placeholder, or else the total value of the number will be changed. To illustrate this, say that you work in a department store and you sold something for \$605. But in writing up the sale, you left out the 0 by mistake and wrote \$65. This would be an expensive mistake; instead of charging 6 hundreds, 0 tens, and 5 ones, you charged 6 tens and 5 ones.

In many division problems, a 0 will be needed in the quotient to serve as a placeholder. An example will show when a zero is needed.

EXAMPLE: Divide 6,122 by 15
Solution:
As you have learned, start the division process by dividing 15 into 61:

```
        4
   15 ⟌6122
4 × 15 = 60 → 60↓
        12 ← new dividend
```

The next step is to divide the new dividend by the divisor. But, note that the new dividend, 12, is smaller than the divisor. So, proceed as follows:

```
        40
   15 ⟌6122
       60
        12
```

Since 15 divides into 12 zero times, place a 0 in the quotient right above the 2 you brought down.

Now, form a new dividend by bringing down the next digit:

```
        40  ← placeholder
   15 ⟌6122
       60↓
        122 ← new dividend
```

Then, divide the new dividend, 122, by the divisor, 15:

```
        408  ← 122 ÷ 15 = 8
   15 ⟌6122
       60
        122
8 × 15 = 120 → 120
          2 ← remainder
```

Finally, state the remainder as a fraction:

$$\begin{array}{r} 408\frac{2}{15} \\ 15\,\overline{)6122} \\ \underline{60} \\ 122 \\ \underline{120} \\ \frac{2}{15} \end{array}$$

Let's now look at another example.

EXAMPLE: Divide 2,472 by 12
Solution:
Start the division process by dividing 12 into 24:

$$\begin{array}{r} 2 \\ 12\,\overline{)2472} \\ \underline{24}\downarrow \\ 7 \end{array} \longleftarrow \text{new dividend}$$

Note that the new dividend, 7, is smaller than the divisor, 12. So, place a 0 in the quotient right above the 7 you brought down:

$$\begin{array}{r} 20 \\ 12\,\overline{)2472} \\ \underline{24} \\ 7 \end{array}$$

Now, bring down the next digit to form a new dividend:

$$\begin{array}{r} 20 \\ 12\,\overline{)2472} \\ \underline{24}\downarrow \\ 72 \end{array} \longleftarrow \text{new dividend}$$

Finally, divide the new dividend, 72, by the divisor, 12:

$$\begin{array}{r} 206 \\ 12\,\overline{)2472} \\ \underline{24} \\ 72 \\ \underline{72} \\ 0 \end{array}$$

72 ÷ 12 = 6 (quotient digit 6)

6 × 12 = 72 (subtracted 72)

Since there is no remainder, the problem "comes out even."

Name ______________________ Date ______________________

UNIT 13 Exercises

Divide each of these problems. Where needed, put zeros in the quotient.

1. $15\overline{)3{,}060}$
2. $24\overline{)2{,}549}$
3. $17\overline{)5{,}219}$
4. $14\overline{)4{,}270}$
5. $18\overline{)9{,}090}$
6. $19\overline{)2{,}052}$
7. $12\overline{)7{,}212}$
8. $27\overline{)2{,}838}$
9. $35\overline{)3{,}815}$
10. $27\overline{)8{,}262}$
11. $13\overline{)5{,}239}$
12. $14\overline{)7{,}014}$

ANSWERS

1. ____________
2. ____________
3. ____________
4. ____________
5. ____________
6. ____________
7. ____________
8. ____________
9. ____________
10. ____________
11. ____________
12. ____________

13. $21\overline{)4{,}368}$ 14. $17\overline{)5{,}440}$ 15. $24\overline{)9{,}672}$ 16. $11\overline{)2{,}279}$

17. $35\overline{)14{,}315}$ 18. $42\overline{)21{,}336}$ 19. $33\overline{)20{,}031}$

20. $25\overline{)10{,}075}$ 21. $18\overline{)10{,}944}$ 22. $17\overline{)15{,}334}$

23. $18\overline{)16{,}218}$ 24. $21\overline{)12{,}789}$ 25. $19\overline{)17{,}195}$

ANSWERS

13. ____________
14. ____________
15. ____________
16. ____________
17. ____________
18. ____________
19. ____________
20. ____________
21. ____________
22. ____________
23. ____________
24. ____________
25. ____________

Name ______________________________ Date ______________________________

26. $39 \overline{)12{,}051}$ 27. $61 \overline{)30{,}805}$ 28. $55 \overline{)16{,}940}$

ANSWERS

26. ____________
27. ____________
28. ____________
29. ____________
30. ____________
31. ____________
32. ____________
33. ____________
34. ____________
35. ____________
36. ____________
37. ____________

29. $81 \overline{)32{,}967}$ 30. $72 \overline{)57{,}888}$ 31. $94 \overline{)19{,}552}$

32. $25 \overline{)50{,}125}$ 33. $17 \overline{)51{,}119}$ 34. $23 \overline{)69{,}046}$

35. $35 \overline{)70{,}175}$ 36. $38 \overline{)228{,}342}$ 37. $61 \overline{)122{,}549}$

38. $52\overline{)416,208}$ 39. $75\overline{)150,525}$ 40. $225\overline{)901,350}$

ANSWERS

38. ____________

39. ____________

40. ____________

41. ____________

42. ____________

43. ____________

44. ____________

41. $335\overline{)1,342,680}$ 42. $724\overline{)2,899,690}$ 43. $882\overline{)2,646,882}$

44. $741\overline{)4,452,669}$

Name ______________________ Date ______________________

UNIT 13 Applications

1. A word processing department can produce 1,632 letters in 8 hours. How many letters can they produce in one hour?

2. A department store sold 525 leather coats during a special 5-day sale. What was the average number of coats sold daily?

3. A local school was given a grant of $83,200 to buy computers. How many computers can the school buy if the price of a computer is $800?

4. A family of 5 won $15,225 on a TV game show. What is each family member's share?

ANSWERS

1. ______________
2. ______________
3. ______________
4. ______________

5. A large industrial machine can make 16,072 units in 8 hours. How many units can it make in one hour?

ANSWERS

5. ____________

6. ____________

7. ____________

8. ____________

6. Sue Teal makes $12,600 a year. What does she make in a month?

7. Angela Kaye's new car gets an average of 35 miles to a gallon of gas. How many gallons of gas will it take her to drive 3,780 miles?

8. A gasoline tank holds 49,500 gallons of gas. How many days would it take to empty the tank if 550 gallons are taken out each day?

Name ______________________ Date ______________________

9. Tuna is packed 24 cans to a case. How many cases will be needed to pack 2,520 cans?

ANSWERS

9. __________

Work Space

UNIT 14

Shortcut Division

In Unit 10, we learned that there are certain shortcuts that can save you time and increase your accuracy when multiplying whole numbers. There are also certain shortcuts that can save you time and increase your accuracy when dividing whole numbers. In this unit, we are going to study two helpful shortcuts to division.

DIVIDING WHEN THE DIVIDEND ENDS WITH ZEROS

When the dividend ends with a zero (or zeros), just move the zero(s) into the quotient and divide by the nonzero whole numbers:

EXAMPLE 1: Divide 4,500 by 9
Solution:

$45 \div 9 = 5 \rightarrow$

$$\begin{array}{r} 500 \\ 9\overline{)4{,}500} \end{array} = 500$$

EXAMPLE 2: Divide 30,000 by 15
Solution:

$30 \div 15 = 2 \rightarrow$

$$\begin{array}{r} 2{,}000 \\ 15\overline{)30{,}000} \end{array} = 2{,}000$$

DIVIDING WHEN THE DIVISOR AND THE DIVIDEND END WITH ZEROS

When **both** the divisor and the dividend end with zeros, drop the zeros at the end of the divisor, and drop the **same** number of zeros at the end of the dividend; then, divide by the nonzero whole numbers.

EXAMPLE 1: Divide 1,200 by 400
Solution:

$12 \div 4 = 3 \rightarrow$

$$\begin{array}{r} 3 \\ 4\not{0}\not{0}\overline{)12\not{0}\not{0}} \\ \underline{12} \end{array} = 3$$

EXAMPLE 2: Divide 45,000 by 500
Solution:

$450 \div 5 = 90 \rightarrow$

$$\begin{array}{r} 90 \\ 5\not{0}\not{0}\overline{)450\not{0}\not{0}} \\ \underline{450} \end{array} = 90$$

EXAMPLE 3: Divide 36,000 by 60
Solution:

$$\begin{array}{r} 600 \\ 6\not{0}\overline{)3600\not{0}} \\ \underline{3600} \end{array} = 600$$

The divisor has only one zero; thus, drop only one zero in the dividend.

EXAMPLE 4: Divide 300,000 by 100,000
Solution:

$$\begin{array}{r} 3 \\ 1\not{0}\not{0},\not{0}\not{0}\not{0}\overline{)3\not{0}\not{0},\not{0}\not{0}\not{0}} \\ \underline{3} \end{array} = 3$$

Work
Space

Name ______________________ Date ______________________

UNIT 14 Exercises

Divide each of these problems.

1. $4\overline{)1{,}600}$ 2. $5\overline{)4{,}500}$ 3. $9\overline{)720}$ 4. $8\overline{)480}$

5. $6\overline{)360}$ 6. $9\overline{)450}$ 7. $5\overline{)250}$ 8. $7\overline{)630}$ 9. $5\overline{)600}$

10. $9\overline{)810}$ 11. $7\overline{)490}$ 12. $8\overline{)6{,}400}$ 13. $7\overline{)4{,}900}$

14. $9\overline{)5{,}400}$ 15. $12\overline{)12{,}000}$ 16. $15\overline{)45{,}000}$ 17. $25\overline{)25{,}000}$

18. $32\overline{)64{,}000}$ 19. $20\overline{)400}$ 20. $40\overline{)800}$ 21. $60\overline{)360}$

22. $20\overline{)200}$ 23. $80\overline{)640}$ 24. $600\overline{)3{,}600}$ 25. $700\overline{)4{,}900}$

ANSWERS

1. ______
2. ______
3. ______
4. ______
5. ______
6. ______
7. ______
8. ______
9. ______
10. ______
11. ______
12. ______
13. ______
14. ______
15. ______
16. ______
17. ______
18. ______
19. ______
20. ______
21. ______
22. ______
23. ______
24. ______
25. ______

26. $300\overline{)1{,}200}$ 27. $800\overline{)64{,}000}$ 28. $200\overline{)14{,}400}$

29. $400\overline{)100{,}000}$ 30. $5{,}000\overline{)25{,}000}$ 31. $5{,}500\overline{)27{,}500}$

32. $6{,}000\overline{)42{,}000}$ 33. $3{,}000\overline{)360{,}000}$ 34. $2{,}000\overline{)2{,}500{,}000}$

35. $45{,}000\overline{)135{,}000}$ 36. $7{,}800\overline{)1{,}950{,}000}$

37. $12{,}000\overline{)1{,}800{,}000}$ 38. $15{,}000\overline{)600{,}000}$

ANSWERS

26. ____________
27. ____________
28. ____________
29. ____________
30. ____________
31. ____________
32. ____________
33. ____________
34. ____________
35. ____________
36. ____________
37. ____________
38. ____________

Name ______________________ Date ______________________

UNIT 14 Applications

Using shortcuts, divide each of these problems. Where possible, do the division in your head.

1. A school district has $60,000 to divide evenly among 200 schools. What is each school's share?

2. How long would it take a plane traveling at 200 miles per hour to fly 3,600 miles?

3. There are around 50,000,000 personal computers in use in the United States. Considering a population of about 230,000,000 people, there is one computer for every ______ people. Please fill in the blank.

4. If Lang King makes $800 for 40 hours of work, what does he make per hour?

5. At a summer camp cookout, 600 hotdogs were made for 200 kids. How many hotdogs were there for each kid?

6. A champion boxer won $2,000,000 for a 10-round fight. What did he earn for each round?

7. At a high school basketball game, $2,000 worth of tickets were sold. If 1,000 people paid to see the game, what was the price for each ticket?

8. A day care center spent $200 on food for 50 children. What was spent for each child?

9. The Herrara Company had a $10,000,000 dividend to be divided among its 5,000,000 shares of stock. What is the dividend per share?

ANSWERS

1. ______________
2. ______________
3. ______________
4. ______________
5. ______________
6. ______________
7. ______________
8. ______________
9. ______________

10. Huron Corporation sold 5,000 units of a certain product for $250,000. What was the selling price per unit?

11. Ray Massengale, a salesperson for a large sporting goods company, sold 1,000 volleyballs to a local school district for $5,000. What was the selling price for each ball?

12. The Gulfside Motel spent $32,000 for cleaning and repairing its 100 rooms. What was the average cost to fix each room?

ANSWERS

10. ____________

11. ____________

12. ____________

UNIT 15 The Meaning of Fractions

Most of the numbers we have worked with so far have been whole numbers. We use whole numbers to represent complete units—such as a gallon of milk, a dollar, a car, a TV set, and a textbook.

There are many times where we need to express less than a complete unit. For example, you often buy a "half" gallon of milk, or you need two "quarters" to put into a soft drink machine, or you may plan to jog "three-quarters" of a mile today. In each of these examples, you are working with values that are less than one whole unit. In this unit, we will start our study of fractions, or partial units.

UNDERSTANDING FRACTIONS

A fraction is used to show that a whole unit has been divided into equal parts. Thus, a **fraction** is a "*part* of a *whole* unit." For example, the box below represents one whole unit (remember that a whole unit is anything that can be counted or measured). But note that the box is divided into four equal parts and that three of the four parts are shaded. This means that we are identifying three of the four parts, or "three-fourths" of the whole box.

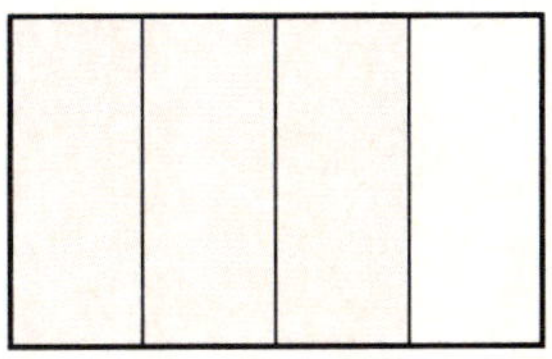

$= \frac{3}{4}$

A fraction, or **common fraction,** is written by separating two numbers by a line, like this: $\frac{3}{4}$. The number that is written under the line shows how many equal parts the unit is divided into; it is called the **denominator.** The number that is written above the line is the **numerator;** it shows how many parts of the whole are expressed by the fraction. The line between the numerator and the denominator is called the **fraction bar.**

3 ⟶ **numerator** = the number of parts expressed by the fraction
–
4 ⟶ **denominator** = the number of parts the unit has been divided into

It is important to remember that the numerator always tells us the number of parts that are in the fraction. And the denominator tells us how many parts the complete unit has been divided into. So, the fraction $\frac{2}{3}$ tells us that a complete unit has been divided into three equal parts, and that two of the three parts are represented by the fraction. The numerator and the denominator are called the **terms** of the fraction.

TYPES OF COMMON FRACTIONS

There are three types of common fractions: proper fractions, improper fractions, and mixed numbers.

Proper Fractions. A **proper fraction** is one with a numerator that is *less* than its denominator. Examples include $\frac{1}{8}$, $\frac{1}{4}$, $\frac{1}{2}$, $\frac{3}{4}$, $\frac{7}{8}$, and $\frac{9}{10}$. A proper fraction expresses less than a whole unit.

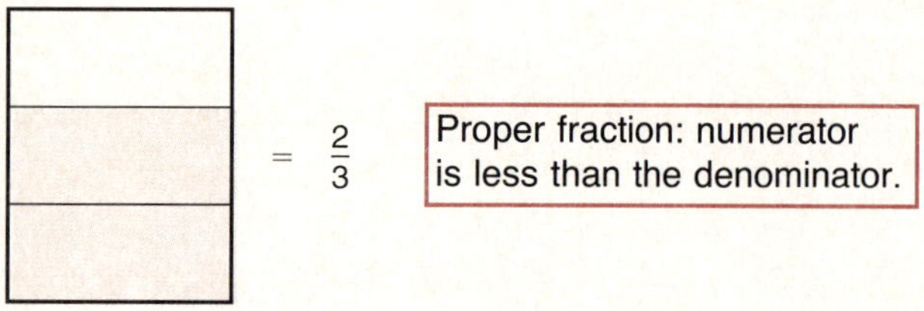

Improper Fractions. An **improper fraction** is one with a numerator that is equal to or greater than its denominator. Examples include $\frac{5}{5}$, $\frac{7}{6}$, $\frac{11}{9}$, $\frac{18}{5}$, $\frac{4}{3}$, and $\frac{19}{15}$. An improper fraction expresses one or more whole units.

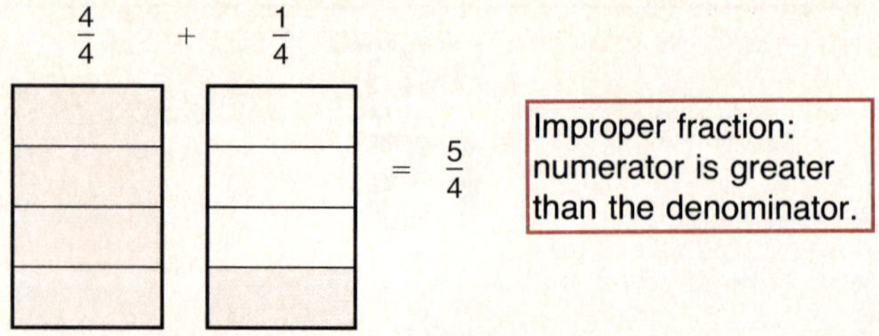

Mixed Numbers. A **mixed number** is a whole number and a common fraction. Examples include $3\frac{1}{2}$, $4\frac{1}{8}$, $5\frac{3}{4}$, and $75\frac{2}{3}$. A mixed number expresses a whole unit and a fraction.

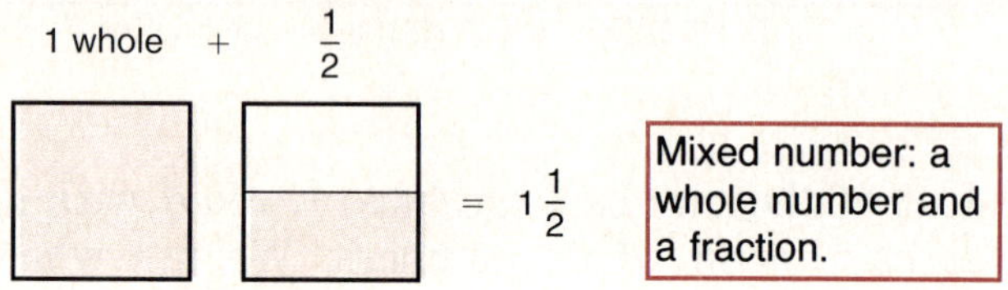

Name ______________________ Date ______________________

UNIT 15 Exercises

Complete these statements.

1. A fraction is part of ______________________.

2. A fraction is written by separating two numbers by ________.

3. The top number in a fraction is called the ______________.

4. The bottom number in a fraction is called the ____________.

5. What does the top number of a fraction tell you? __________ ____.

6. What does the bottom number of a fraction tell you? ________ ____.

7. In the fraction $\frac{6}{7}$, how many parts are being expressed by the fraction?

8. In the fraction $\frac{11}{12}$, how many equal parts has the whole been divided into?

9. A fraction whose top number is less than its bottom number is called a(n) __________ fraction.

10. A fraction whose top number is equal to or greater than its bottom number is called a(n) __________ fraction.

11. A ________________ combines a whole number and a fraction.

12. If one complete unit is divided into 6 equal parts and 2 are expressed, what will be the denominator of the fraction?

13. Express each of the following as a fraction:
    a. three of four parts

    b. three of seven parts

    c. four of nine parts

    d. eleven of seventeen parts

    e. six of thirteen parts

    f. fifteen of thirty-one parts

    g. thirty-three of fifty-eight parts

    h. ninety-one of one hundred parts

14. What fraction is shown by each of these pictures?

a.

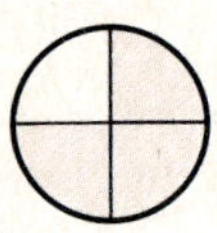

b.

c.

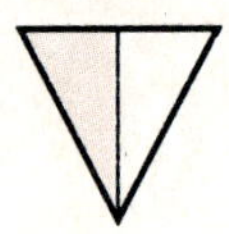

d.

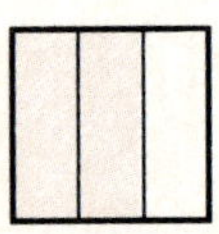

e.

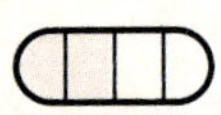

f.

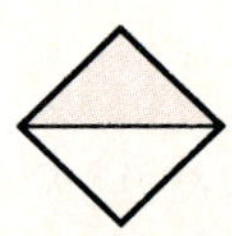

Name ______________________ Date ______________________

UNIT 15 Applications

1. A high school football team won 8 out of 12 football games. What fraction of their games did they win?

2. A business math class has 10 girls and 20 boys. What fraction of the class is made up of girls?

3. A quarterback completed 17 passes in 21 attempts. What fraction of the passes did he complete?

4. Out of a class of 45 students, 12 were sick. What fraction of the class was sick?

5. In May, Greg went on a diet to lose 15 pounds. By July he had lost 11 pounds. What fraction of his weight loss goal has Greg achieved?

ANSWERS

1. ______

2. ______

3. ______

4. ______

5. ______

6. Twila bought a six-pack of cola to take to the beach. By 12 o'clock, Twila had finished 4 cans. What fraction of the colas is left for the afternoon?

7. Conrad bought a 32-ounce container of milk and used 8 ounces to make a milkshake. What fraction of the milk did Conrad use for the shake?

8. Out of 60 questions on a history test, a student marked 52 correct answers. What fraction of the test did the student get correct?

9. Kim's weekly paycheck is $225. Of this amount, Kim saves $75. What fraction of her paycheck does she save?

10. The price of a new car is $12,000, but you must make a down payment of $2,000. The down payment is what fraction of the price of the car?

ANSWERS

6. __________

7. __________

8. __________

9. __________

10. __________

UNIT

16

Reducing Fractions to Lowest Terms

Let's say you ordered pizza when pizzas were on a "two for one" special. When the pizzas were delivered, you cut one into four equal parts, and you cut the second pizza into two equal parts. You now have two pizzas, each divided into equal parts. One pizza has four equal parts, and the other pizza has two equal parts. If you ate two pieces of the pizza cut into four parts, you would have eaten $\frac{2}{4}$ of the pizza. But, if you ate one piece of the pizza cut into two parts, you would have eaten $\frac{1}{2}$ of the pizza.

Either way, you would have eaten the same amount of pizza, because 2 parts out of 4 is the same amount as 1 part out of 2. In other words, $\frac{2}{4} = \frac{1}{2}$. Fractions that represent the same value are called **equivalent fractions.**

$\frac{2}{4} = \frac{1}{2}$

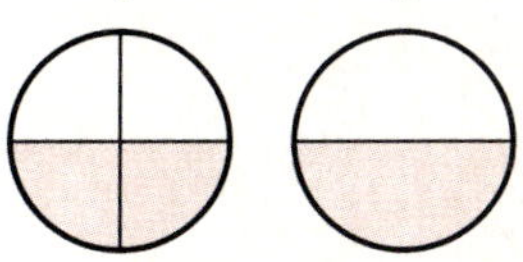

$\frac{2}{4}$ and $\frac{1}{2}$ are equivalent fractions because they have the same value.

REDUCING FRACTIONS TO LOWEST TERMS

Fractions with smaller terms—remember the terms of a fraction are the numerator and the denominator—are easier to work with than fractions with larger terms. So, answers to problems dealing with fractions should be reduced to **lowest terms.** To reduce a fraction to lowest terms, select any number (other than 1) that will divide evenly into *both* the numerator and the denominator. Keep dividing until the terms can no longer be evenly divided by the same number. The following rule will help you when reducing fractions:

RULE

Dividing both terms of a fraction by the same number does not change the value of the fraction.

EXAMPLE 1: Reduce $\frac{6}{8}$ to the lowest terms

Solution:

Select a number that will divide evenly into both the numerator and the denominator. By simple observation we can see that 2 will divide evenly into both terms of this fraction.

$$\frac{6}{8} = \frac{6 \div 2}{8 \div 2} = \frac{3}{4}$$

We have now reduced $\frac{6}{8}$ to $\frac{3}{4}$. The fraction is now in its lowest terms because there is no number that can be divided evenly into both the 3 and the 4.

EXAMPLE 2: Reduce $\frac{25}{75}$ to lowest terms.

Solution:

Select a number that will divide evenly into the terms of the fraction. Let's use 5.

$$\frac{25}{75} = \frac{25 \div 5}{75 \div 5} = \frac{5}{15}$$

We have now reduced $\frac{25}{75}$ to $\frac{5}{15}$. But is the fraction in lowest terms? No. Because $\frac{5}{15}$ can be reduced again. Let's finish reducing it by again dividing both terms by 5:

$$\frac{5}{15} = \frac{5 \div 5}{15 \div 5} = \frac{1}{3}$$

Not all fractions can be reduced. If there is no number that will divide evenly into both the numerator and the denominator, the fraction is already in its lowest terms.

EXAMPLE: Reduce $\frac{13}{17}$ to lowest terms.

Solution:

There is no number that will divide evenly into both the numerator and the denominator. Thus, the terms of the fraction cannot be reduced.

Name ______________________ Date ______________________

UNIT 16 Exercises

Where possible, reduce each of these fractions to lowest terms.

1. $\frac{6}{12}$ 2. $\frac{4}{8}$ 3. $\frac{3}{4}$ 4. $\frac{6}{9}$ 5. $\frac{5}{8}$

6. $\frac{9}{12}$ 7. $\frac{4}{16}$ 8. $\frac{8}{12}$ 9. $\frac{11}{12}$ 10. $\frac{16}{24}$

11. $\frac{8}{16}$ 12. $\frac{7}{14}$ 13. $\frac{21}{33}$ 14. $\frac{18}{45}$ 15. $\frac{12}{36}$

16. $\frac{22}{26}$ 17. $\frac{34}{66}$ 18. $\frac{12}{80}$ 19. $\frac{72}{81}$ 20. $\frac{14}{28}$

21. $\frac{56}{92}$ 22. $\frac{44}{80}$ 23. $\frac{36}{63}$ 24. $\frac{45}{50}$ 25. $\frac{10}{50}$

26. $\frac{40}{90}$ 27. $\frac{45}{90}$ 28. $\frac{35}{45}$ 29. $\frac{21}{63}$ 30. $\frac{16}{54}$

ANSWERS

1. ________
2. ________
3. ________
4. ________
5. ________
6. ________
7. ________
8. ________
9. ________
10. ________
11. ________
12. ________
13. ________
14. ________
15. ________
16. ________
17. ________
18. ________
19. ________
20. ________
21. ________
22. ________
23. ________
24. ________
25. ________
26. ________
27. ________
28. ________
29. ________
30. ________

31. $\frac{22}{44}$ 32. $\frac{25}{75}$ 33. $\frac{36}{36}$ 34. $\frac{72}{96}$ 35. $\frac{90}{95}$

36. $\frac{120}{240}$ 37. $\frac{50}{150}$ 38. $\frac{75}{300}$ 39. $\frac{280}{340}$

40. $\frac{300}{600}$ 41. $\frac{450}{500}$ 42. $\frac{135}{725}$ 43. $\frac{720}{900}$

44. $\frac{400}{4000}$ 45. $\frac{5,000}{50,000}$

ANSWERS

31. ____________
32. ____________
33. ____________
34. ____________
35. ____________
36. ____________
37. ____________
38. ____________
39. ____________
40. ____________
41. ____________
42. ____________
43. ____________
44. ____________
45. ____________

Name ______________________ Date ______________________

UNIT 16 Applications

In these problems, reduce each final answer to lowest terms, if needed.

1. Maureen O'Shea can jog 4 miles. Her goal is to be able to jog 6 miles. What fraction of her goal has she reached?

2. Bill Seymour's gross pay was $200, but $40 was taken out in taxes. What fraction of Bill's gross pay was taken out in taxes?

3. In a clearance sale, a department store sold 50 fur coats. What fraction of fur coats was sold if the original number in stock was 70?

4. Mona agreed to sell 20 boxes of cookies for her school. At the end of the first week of the sale, Mona had sold 12 boxes. What fraction does she have left to sell?

5. Of a graduating class of 600, there were 350 girls. What fraction of the class was girls?

6. A coat was marked down from $75 to $50. The savings is what fraction of the original price?

7. Evan Bonert bought a new home for $80,000 and made a down payment of $8,000. What fraction of the original price was the down payment?

8. Wendy Rosenburg jogged 20 minutes out of an hour she was on the track. (a) What fraction of the hour did Wendy spend jogging? (b) What fraction of the hour did Wendy spend walking?

ANSWERS

1. ____________
2. ____________
3. ____________
4. ____________
5. ____________
6. ____________
7. ____________
8. a. ____________
   b. ____________

Work Space

UNIT 17 Determining the Greatest Common Divisor

In Unit 16, you learned that a fraction can be reduced to *lower terms* by dividing its terms (the numerator and the denominator) by the same number. A fraction is reduced to *lowest terms* by dividing its terms by the same number until further division is not possible. Let's take an example as a quick review.

EXAMPLE: Reduce $\frac{16}{96}$ to lowest terms:
Solution:
Select a number that will divide evenly into both 12 and 96. Let's use 4:

$$\frac{16}{96} = \frac{16 \div 4}{96 \div 4} = \frac{4}{24}$$

Now, we must reduce $\frac{4}{24}$ again because it is not in its lowest terms. Let's divide by 4 again:

$$\frac{4}{96} = \frac{4 \div 4}{24 \div 4} = \frac{1}{6}$$

To reduce this fraction, we selected a number (4) that would divide evenly into both terms of the fraction. We kept dividing until the terms of the fraction could no longer be divided evenly by the same number. Basically, we are using a method of trial and error. We select a number and see if it will divide evenly into both terms of the fraction. If it does not, we select another number and try again. This method works well for many fractions. However, it can be time-consuming if numbers that are common to the numerator and the denominator are not obvious.

For example, how would you reduce the fraction $\frac{39}{65}$? You can't use 3 because only the numerator (39) is evenly divisible by 3; the denominator (65) is not. You can't use 2 because neither term is evenly divisible by 2. So, what number do you use?

You can use the following method to determine the **greatest common divisor** (GCD), that is, the largest number that will divide evenly into both the numerator and the denominator.

Step 1: Divide the numerator into the denominator. If it divides evenly without a remainder, the numerator is the greatest common divisor. If a remainder exists, go on to Step 2.

Step 2: Divide the remainder into the divisor from Step 1.

Step 3: Keep dividing the remainders into the previous divisors until there is no remainder. The *last divisor* will be your greatest common divisor.

Note: If you obtain a remainder of 1, it means that the fraction is already in its lowest terms.

Now, let's find the greatest common divisor and reduce $\frac{39}{65}$ to lowest terms:

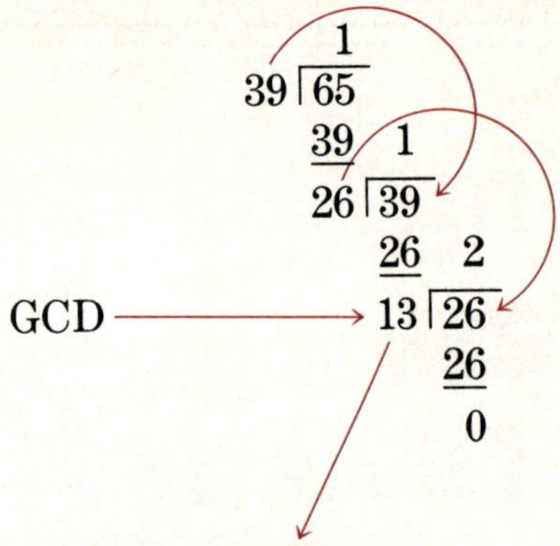

$$\frac{39}{65} = \frac{39 \div 13}{65 \div 13} = \frac{3}{5}$$

Even though you can use any common divisor to reduce a fraction, you should always use the greatest common divisor because the greatest common divisor will reduce the fraction to lowest terms in one step.

Name ______________________ Date ______________________

UNIT 17 Exercises

Find the greatest common divisor and reduce these fractions to lowest terms.

1. $\frac{35}{75}$
2. $\frac{45}{135}$
3. $\frac{30}{92}$
4. $\frac{75}{225}$
5. $\frac{68}{238}$
6. $\frac{72}{192}$
7. $\frac{22}{324}$
8. $\frac{55}{255}$
9. $\frac{234}{288}$
10. $\frac{350}{725}$
11. $\frac{675}{900}$
12. $\frac{250}{550}$
13. $\frac{430}{920}$
14. $\frac{750}{1,000}$
15. $\frac{540}{1,000}$
16. $\frac{270}{1,200}$
17. $\frac{250}{1,500}$
18. $\frac{300}{1,200}$
19. $\frac{525}{2,500}$
20. $\frac{450}{1,600}$
21. $\frac{4,000}{8,000}$
22. $\frac{3,225}{6,000}$
23. $\frac{2,550}{8,500}$
24. $\frac{900}{9,050}$
25. $\frac{3,200}{8,000}$

ANSWERS

1. ______
2. ______
3. ______
4. ______
5. ______
6. ______
7. ______
8. ______
9. ______
10. ______
11. ______
12. ______
13. ______
14. ______
15. ______
16. ______
17. ______
18. ______
19. ______
20. ______
21. ______
22. ______
23. ______
24. ______
25. ______

Work Space

Name ______________________ Date ______________________

UNIT 17 Applications

In each of these problems, give the final answer in lowest terms.

1. What fraction of an hour is
   a. 15 minutes
   b. 30 minutes
   c. 45 minutes

2. Terry Applebee, a salesperson, visited 36 customers and made sales to 12 of them. What fraction of her customers did she sell to?

3. Last week, Tim Yates earned \$375, but his employer withheld \$50 for taxes. What fraction of Tim's pay was withheld?

4. A new house is for sale at \$80,000 with a required \$16,000 down payment. The down payment is what fraction of the price of the house?

ANSWERS

1. a. ________
   b. ________
   c. ________
2. ________
3. ________
4. ________

5. An item marked for sale at $425 is being sold for $300. The savings is what fraction of the original price?

ANSWERS

5. ____________
6. ____________
7. ____________
8. ____________

6. Of the 2,200 employees of Hillsdale Company, 950 are women. What fraction of the staff is made up of women?

7. During a clearance sale, Windsor Men's Shop sold 148 blazers. What fraction of the total stock was sold if there were 240 blazers in stock before the sale?

8. Out of a total graduating class of 650 students, 525 went on to college. What fraction of the graduating class *did not* go to college?

UNIT 18

Raising Fractions to Higher Terms (Higher Equivalents)

In Units 16 and 17, we worked with equivalent fractions, that is, fractions that have the same value. We learned that a fraction can be reduced to lowest terms by dividing both the numerator and the denominator by the same number until further division is not possible.

Even though final answers dealing with fractions are usually reduced to lowest terms, there are some operations with fractions in which you will need to *raise* a fraction to higher terms. For example, to add and subtract fractions with different denominators it is usually necessary to raise a fraction to higher terms.

A fraction is raised to higher terms by multiplying *both* terms of the fraction by the *same* number. The following rule will help you when raising fractions to higher terms:

RULE

> Multiplying both terms of a fraction by the same number does not change the value of the fraction.

For example, we can raise $\frac{1}{2}$ to higher terms by multiplying both the numerator and the denominator by any number we choose. Let's use 3 for this example:

$$\frac{1}{2} = \frac{1 \times 3}{2 \times 3} = \frac{3}{6}$$

We have now raised $\frac{1}{2}$ to $\frac{3}{6}$, which is a higher equivalent. In effect, we have just renamed the fraction, because $\frac{1}{2}$ and $\frac{3}{6}$ have the same value; they are equivalent.

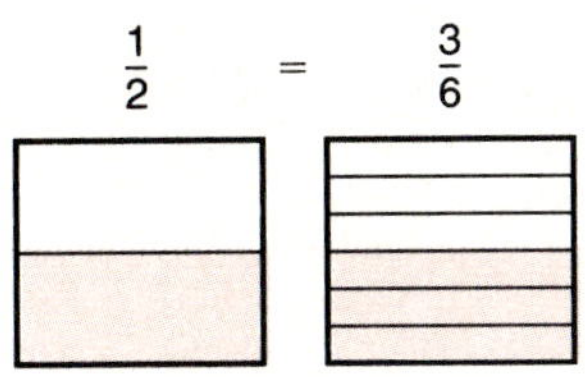

$\frac{1}{2}$ and $\frac{3}{6}$ are equivalent fractions because they represent the same value.

Because $\frac{3}{3} = 1$, multiplying or dividing by such a number does not change the value of the fraction because the fraction is actually being multiplied or divided by 1.

RAISING A FRACTION TO HIGHER TERMS WHEN A NEW DENOMINATOR IS ALREADY GIVEN

Sometimes you will need to raise a fraction to higher terms when a new denominator (the higher term) is already given. For example,

$$\frac{3}{4} = \frac{?}{16}$$

To raise such a fraction to higher terms, follow these steps:

Step 1: Divide the original denominator into the new denominator.

Step 2: Multiply the numerator by the quotient you found in Step 1, and write the result over the new denominator.

To continue the above example, let's now raise $\frac{3}{4}$ to a higher equivalent with a denominator of 16.

Step 1: Divide the original denominator, 4, into the new denominator, 16:

$$\frac{3}{4} = \frac{?}{16} \rightarrow 4\overline{)16}^{\;4}$$

Step 2: Multiply the numerator by the 4 you obtained in Step 1 (4 × 3 = 12), and place the result over the new denominator:

$$\boxed{4 \times 3 = 12} \rightarrow \frac{3}{4} = \frac{12}{16}$$

We know that our answer is correct because:

$$\frac{4}{4} \times \frac{3}{4} = \frac{12}{16}$$

Remember, the value of the fraction is not changed; we only renamed the fraction because we multiplied by $\frac{4}{4}$, which is equal to 1.

Let's look at another example.

EXAMPLE: Raise the following fraction as indicated:

$$\frac{5}{7} = \frac{?}{42}$$

Solution:

Step 1: Divide the original denominator, 7, into the new denominator, 42:

$$7\overline{)42}^{\;6}$$

Step 2: Multiply the numerator, 5, by the 6 you obtained in Step 1 (6 × 5 = 30), and place the result over the original denominator.

$$\boxed{6 \times 5 = 30} \rightarrow \frac{5}{7} = \frac{30}{42}$$

Name ______________________ Date ______________________

UNIT 18 Exercises

Raise these fractions to the term indicated.

1. $\frac{3}{4} = \frac{?}{8}$
2. $\frac{2}{3} = \frac{?}{21}$
3. $\frac{3}{5} = \frac{?}{15}$
4. $\frac{2}{7} = \frac{?}{28}$
5. $\frac{4}{5} = \frac{?}{20}$
6. $\frac{7}{9} = \frac{?}{45}$
7. $\frac{5}{8} = \frac{?}{40}$
8. $\frac{3}{5} = \frac{?}{15}$
9. $\frac{5}{6} = \frac{?}{30}$
10. $\frac{3}{7} = \frac{?}{21}$
11. $\frac{7}{8} = \frac{?}{40}$
12. $\frac{2}{3} = \frac{?}{24}$
13. $\frac{8}{9} = \frac{?}{72}$
14. $\frac{7}{8} = \frac{?}{64}$
15. $\frac{2}{7} = \frac{?}{28}$
16. $\frac{8}{12} = \frac{?}{36}$

ANSWERS

1. ____________
2. ____________
3. ____________
4. ____________
5. ____________
6. ____________
7. ____________
8. ____________
9. ____________
10. ____________
11. ____________
12. ____________
13. ____________
14. ____________
15. ____________
16. ____________

Raise each of these fractions as indicated.

ANSWERS

17. $\frac{1}{4}$ to 16ths
18. $\frac{3}{8}$ to 32nds
19. $\frac{2}{15}$ to 45ths
20. $\frac{5}{9}$ to 72nds

21. $\frac{9}{12}$ to 48ths
22. $\frac{14}{15}$ to 75ths
23. $\frac{32}{40}$ to 80ths
24. $\frac{17}{32}$ to 64ths

25. $\frac{12}{18}$ to 36ths

17. ________
18. ________
19. ________
20. ________
21. ________
22. ________
23. ________
24. ________
25. ________

Name ______________________ Date ______________________

UNIT 18 Applications

Use equivalent fractions to solve the following application problems.

ANSWERS

1. a. ________
   b. ________
   c. ________
   d. ________
2. ________
3. ________

1. Change each of these amounts as indicated:

   a. $\frac{3}{4}$ of a quart to ounces (32nds)

   b. $\frac{1}{8}$ of a pound to ounces (16ths)

   c. $\frac{1}{4}$ of a foot to inches (12ths)

   d. $\frac{1}{2}$ of a gallon to quarts (4ths)

2. If 4 out of 5 fourth graders are going on a picnic, how many of the 40 students in the class will be going?

3. In a class of 24 students, $\frac{2}{3}$ are boys. How many girls are in the class?

4. What would be the new selling price of a coat that had cost $72 and was marked down by $\frac{1}{4}$?

5. Sam and Julie ate $\frac{4}{5}$ of a large pizza after cutting it into 10 equal pieces. How many pieces did they eat?

ANSWERS

4. ____________

5. ____________

UNIT 19 Improper Fractions and Mixed Numbers

You know that a fraction is used to express a part of a whole unit, or something less than one. There are times, however, when fractions are used to express more than a unit. For example, you know that a quarter is $\frac{1}{4}$ of a dollar (a quarter can be written as $\$\frac{1}{4}$). If you had 3 quarters, you would have $\frac{3}{4}$ of a dollar. But what if you had 5 quarters? Then you would have $\frac{5}{4}$ of a dollar, which means that you have one dollar ($\frac{4}{4}$) plus $\frac{1}{4}$ of another dollar.

$\frac{4}{4}$ and $\frac{1}{4}$ = $\frac{5}{4}$

We call $\frac{5}{4}$ improper because its numerator is larger than its denominator (remember that an improper fraction is one whose numerator is equal to or greater than its denominator). Other examples of improper fractions are $\frac{6}{5}$, $\frac{8}{8}$, $\frac{11}{9}$, $\frac{21}{10}$, and $\frac{75}{5}$.

MIXED NUMBERS

A mixed number is made up of a whole number and a fraction. Examples of mixed numbers include $2\frac{1}{2}$, $3\frac{1}{3}$, $5\frac{3}{4}$, and $8\frac{6}{7}$. A mixed number represents a value that is equal to or greater than one whole unit. If you lost $5\frac{1}{2}$ pounds, for example, you would have lost five whole pounds, plus $\frac{1}{2}$ of another pound.

All mixed numbers can be expressed as improper fractions, and all improper fractions can be expressed as mixed numbers. Let's look at how it's done.

CONVERTING MIXED NUMBERS TO IMPROPER FRACTIONS

To convert (to change) a mixed number to an improper fraction, follow these steps:

Step 1: Multiply the whole number by the denominator of the fraction.

Step 2: Add the product you obtained in Step 1 to the numerator of the fraction.

Step 3: Write the total from Step 2 over the original denominator.

EXAMPLE 1: Convert $6\frac{1}{3}$ to an improper fraction.

Solution:

Step 1: Multiply the whole number, 6, by the denominator of the fraction, 3:

$$6\frac{1}{3} \rightarrow 3 \times 6 = 18$$

Step 2: Add the product from Step 1 (18) to the numerator of the fraction 1:

$$18 + 1 = 19$$

Step 3: Write the total from Step 2 (19) over the original denominator:

$$6\frac{1}{3} = (3 \times 6 = 18) + 1 = \frac{19}{3}$$

EXAMPLE 2: Convert $6\frac{7}{8}$ to an improper fraction.

Solution:

$$6\frac{7}{8} = (8 \times 6 = 48) + 7 = \frac{55}{8}$$

CONVERTING IMPROPER FRACTIONS TO WHOLE OR MIXED NUMBERS

To convert an improper fraction to a whole or mixed number, you divide the denominator into the numerator. If there is no remainder, the answer is a whole number. If there is a remainder, the improper fraction is a mixed number; state the remainder as a fraction over the original denominator.

EXAMPLE 1: Convert $\frac{45}{9}$ to a whole or mixed number.
Solution:

$$\begin{array}{r} 5 \\ 9\overline{)45} \end{array} = 5$$

Since the problem came out even (no remainder), the answer is a whole number.

EXAMPLE 2: Convert $\frac{21}{5}$ to a whole or mixed number.
Solution:

$$\begin{array}{r} 4 \\ 5\overline{)21} \\ \underline{20} \\ 1 \end{array} = 4\tfrac{1}{5}$$

remainder → 1

Since the problem had a remainder, the answer is a mixed number.

EXAMPLE 3: Convert $\frac{55}{8}$ to a whole or mixed number.
Solution:

$$\begin{array}{r} 6 \\ 8\overline{)55} \\ \underline{48} \\ 7 \end{array} = 6\tfrac{7}{8}$$

Name ______________________ Date ______________________

UNIT 19 Exercises

Change each of these mixed numbers to improper fractions.

1. $3\frac{3}{4}$ 2. $5\frac{1}{4}$ 3. $6\frac{7}{8}$ 4. $12\frac{1}{8}$ 5. $15\frac{1}{3}$ 6. $22\frac{1}{2}$ 7. $9\frac{2}{3}$

8. $15\frac{1}{2}$ 9. $6\frac{1}{15}$ 10. $7\frac{4}{5}$ 11. $10\frac{1}{4}$ 12. $21\frac{2}{3}$ 13. $5\frac{1}{9}$ 14. $3\frac{5}{6}$

15. $14\frac{5}{6}$ 16. $16\frac{4}{5}$

Change each of these improper fractions to a mixed number or to a whole number. Reduce the fractional part of your answers.

17. $\frac{12}{5}$ 18. $\frac{13}{7}$ 19. $\frac{34}{6}$ 20. $\frac{11}{8}$ 21. $\frac{9}{2}$ 22. $\frac{15}{3}$ 23. $\frac{17}{4}$

24. $\frac{17}{8}$ 25. $\frac{39}{5}$ 26. $\frac{31}{7}$ 27. $\frac{25}{3}$ 28. $\frac{47}{4}$ 29. $\frac{52}{7}$ 30. $\frac{15}{2}$

ANSWERS

1. ____________
2. ____________
3. ____________
4. ____________
5. ____________
6. ____________
7. ____________
8. ____________
9. ____________
10. ____________
11. ____________
12. ____________
13. ____________
14. ____________
15. ____________
16. ____________
17. ____________
18. ____________
19. ____________
20. ____________
21. ____________
22. ____________
23. ____________
24. ____________
25. ____________
26. ____________
27. ____________
28. ____________
29. ____________
30. ____________

Work Space

Name ______________________ Date ______________________

UNIT 19 Applications

1. How much money does each of these improper fractions represent?

   a. $\frac{\$12}{4} =$

   b. $\frac{\$15}{5} =$

   c. $\frac{\$21}{5} =$

   d. $\frac{\$18}{6} =$

   e. $\frac{\$33}{10} =$

2. Write 37 months in terms of years.

3. Write 12 quarters in terms of dollars. (Hint: one quarter equals $\frac{1}{4}$ of a dollar.)

4. A play was on Broadway for 32 months. How many years was that?

ANSWERS

1. a. ________
   b. ________
   c. ________
   d. ________
   e. ________
2. ________
3. ________
4. ________

5. Write these as improper fractions.

a. $\$12\frac{1}{4}$

b. $\$6\frac{7}{8}$

c. $\$5\frac{3}{4}$

d. $\$14\frac{1}{3}$

ANSWERS

5. a. ________

b. ________

c. ________

d. ________

UNIT 20 Adding Fractions with the Same Denominator

Fractions must have the same denominator to be added. If you are working with two or more fractions that already have the same denominator (or common denominator), the addition process is very simple: add the numerators and place the sum over the common denominator; reduce the final answer to lowest terms, if needed.

EXAMPLE: Add these fractions.

$$\frac{2}{15} + \frac{7}{15}$$

Solution:

Step 1: Add the numerators:

$$2 + 7 = 9$$

Step 2: Place the sum of the numerators, 9, over the common denominator, 15:

$$\frac{2}{15} + \frac{7}{15} = \frac{9}{15}$$

Step 3: Reduce answer to lowest terms (remember, to reduce a fraction, find a number that will divide evenly into both the numerator and the denominator; the number needed to reduce this fraction is 3).

$$\frac{9 \div 3}{15 \div 3} = \frac{3}{5}$$

If you add a set of fractions and get an improper fraction as your sum, you should convert it to a whole number or a mixed number.

EXAMPLE: Add: $\frac{3}{5} + \frac{2}{5} + \frac{4}{5}$

Solution:

Step 1: Add the numerators:

$$3 + 2 + 4 = 9$$

Step 2: Place the sum of the numerators, 9, over the common denominator, 5:

$$\frac{3}{5} + \frac{2}{5} + \frac{4}{5} = \frac{9}{5}$$

Step 3: Convert the improper fraction to a mixed number:

$$\frac{9}{5} \quad = \quad \begin{array}{r} 1 \\ 5\overline{)9} \\ \underline{5} \\ 4 \end{array} \quad = \quad 1\frac{4}{5}$$

Remember. When adding fractions with the same denominator, *add only the numerators.*

Work Space

Name ______________________ Date ______________________

UNIT 20 Exercises

Add these fractions. Change improper fractions to whole or mixed numbers and reduce all final answers to lowest terms.

1. $\frac{1}{3} + \frac{2}{3}$
2. $\frac{3}{8} + \frac{5}{8}$
3. $\frac{1}{5} + \frac{3}{5}$
4. $\frac{2}{9} + \frac{6}{9}$
5. $\frac{3}{7} + \frac{4}{7} + \frac{5}{7}$
6. $\frac{1}{8} + \frac{3}{8} + \frac{7}{8}$
7. $\frac{2}{15} + \frac{4}{15} + \frac{7}{15}$
8. $\frac{3}{11} + \frac{4}{11} + \frac{6}{11}$
9. $\frac{5}{12} + \frac{7}{12} + \frac{11}{12}$
10. $\frac{3}{16} + \frac{9}{16} + \frac{1}{16}$
11. $\frac{8}{25} + \frac{11}{25} + \frac{16}{25}$
12. $\frac{5}{21} + \frac{8}{21} + \frac{11}{21}$

ANSWERS

1. ______
2. ______
3. ______
4. ______
5. ______
6. ______
7. ______
8. ______
9. ______
10. ______
11. ______
12. ______

13. $\frac{7}{32} + \frac{15}{32} + \frac{17}{32}$

14. $\frac{9}{13} + \frac{8}{13} + \frac{12}{13}$

15. $\frac{19}{20} + \frac{17}{20} + \frac{20}{20}$

16. $\frac{12}{17} + \frac{14}{17} + \frac{6}{17}$

17. $\frac{19}{21} + \frac{16}{21} + \frac{10}{21} + \frac{8}{21}$

18. $\frac{10}{25} + \frac{12}{25} + \frac{14}{25} + \frac{7}{25}$

19. $\frac{25}{40} + \frac{13}{40} + \frac{11}{40} + \frac{9}{40}$

20. $\frac{8}{35} + \frac{15}{35} + \frac{12}{35} + \frac{7}{35}$

21. $\begin{array}{r} \frac{12}{15} \\ \frac{11}{15} \\ + \frac{8}{15} \\ \hline \end{array}$

22. $\begin{array}{r} \frac{6}{42} \\ \frac{3}{42} \\ + \frac{15}{42} \\ \hline \end{array}$

23. $\begin{array}{r} \frac{11}{30} \\ \frac{7}{30} \\ + \frac{17}{30} \\ \hline \end{array}$

ANSWERS

13. ________
14. ________
15. ________
16. ________
17. ________
18. ________
19. ________
20. ________
21. ________
22. ________
23. ________

Name ______________________ Date ______________________

UNIT 20 Applications

ANSWERS

1. ________
2. ________
3. ________
4. ________
5. ________

1. A large pizza was cut into 8 pieces. Bill ate 2 pieces, Jane ate 2 pieces, Sue ate 1 piece, and Tim ate 2 pieces. What fraction of the pizza was eaten?

2. A recipe for bran muffins calls for $\frac{1}{2}$ cup whole wheat flour, $\frac{1}{2}$ cup wheat bran, $\frac{1}{2}$ cup honey, and $\frac{1}{2}$ cup vegetable oil. In total, the recipe calls for how many cups?

3. Dave cut $\frac{3}{4}$ inch off a shelf to get it to fit in a bookcase. When the shelf still didn't fit, Dave cut another $\frac{1}{4}$ inch off. What was the total amount Dave cut off?

4. A pattern calls for the following material: $\frac{7}{8}$ foot of cotton fabric, $\frac{3}{8}$ foot of wool fabric, and $\frac{1}{8}$ foot of lace. How many feet of fabric does the pattern call for?

5. Don worked $\frac{5}{8}$ of an hour overtime on Monday, $\frac{3}{8}$ of an hour overtime on Wednesday, and $\frac{7}{8}$ of an hour overtime on Friday. How many hours overtime did Don work for the week?

6. A recipe for trail mix calls for 8 ounces of sunflower seeds, 8 ounces of pumpkin seeds, 8 ounces of walnuts, 8 ounces of cashews, 8 ounces of almonds, and 2 ounces of coconut shreds. How many pounds of trail mix would the recipe make? (Remember, 1 ounce = $\frac{1}{16}$ of a pound)

ANSWERS

6. ____________
7. ____________
8. ____________

7. A hotdog stand used these numbers of buns during the homecoming game of a football team: 10, 18, 24, 36. How many dozen buns were used? (Remember, one bun = $\frac{1}{12}$ of a dozen.)

8. Tonya sold the following lengths of fabric: 2 feet, 3 feet, 8 feet, and 5 feet. How many yards of fabric did Tonya sell? (One foot equals $\frac{1}{3}$ of a yard.)

UNIT

21 Adding Fractions with Different Denominators

In Unit 20, we added groups of fractions that had the same denominator. When all fractions in a group have the same denominator, we say that they have a "common denominator."

In many operations dealing with fractions, you will have to add fractions that do not have the same denominator. To add a group of fractions with *unlike* denominators, you must change each fraction to an equivalent fraction with a common denominator.

To find a common denominator, you find a number into which all denominators will evenly divide. This can often be done by simple observation.

EXAMPLE: Add: $\frac{1}{2} + \frac{1}{3}$

Solution: Since halves and thirds are not common terms, we must find a common denominator. By simple observation, we can see that both denominators will divide evenly into 6. Thus, 6 is a common denominator. We now raise each fraction to 6ths and add the numerators:

$$\begin{array}{rcl} \frac{1}{2} = \frac{?}{6} & & \frac{1}{2} = \frac{3}{6} \\ + \quad \frac{1}{3} = \frac{?}{6} & = & \frac{1}{3} = \frac{2}{6} \\ \hline & & \frac{5}{6} \end{array}$$

Remember: divide the new denominator, 6, by the old denominator, 2; then multiply the result, 3, by the numerator, 1:

6 ÷ 2 = 3; 3 × 1 = 3

6 ÷ 3 = 2; 2 × 1 = 2

Since a set of fractions can have many common denominators, you can make the problem easier by using the **least common denominator** (LCD). The least common denominator is the smallest number that all denominators will evenly divide into. In the example above, for example, the LCD is 6, because 6 is the smallest number that both 2 and 3 will divide evenly.

FINDING THE LCD USING THE PRIME NUMBER METHOD

When a common denominator cannot be determined by simple observation, you can find the LCD by using the **prime number method.** A **prime number** is a number that can be divided evenly by itself and one, and no other number. Examples of prime numbers are 2, 3, 5, 7, 11, 13, 17, 19, and 23. To illustrate how to find the LCD using the prime number method, let's look at the next example.

EXAMPLE: Find the LCD and add: $\frac{1}{6} + \frac{5}{12} + \frac{4}{15}$

Solution:

Step 1: Write the denominators in a row, like this:

$$\overline{)6 - 12 - 15}$$

Step 2: Choose a prime number that will divide evenly into as many denominators as possible. For this example, let's start with 3:

$$\begin{array}{r} 3\overline{)6 - 12 - 15} \\ 2 - 4 - 5 \end{array}$$

Step 3: Choose another prime number that will divide evenly into as many numbers as possible in the second row. Let's use 2, because it will divide evenly into the 2 and the 4:

$$\begin{array}{l} 3\overline{)6 - 12 - 15} \\ 2\overline{)2 - 4 - 5} \\ 1 - 2 - 5 \end{array}$$

Note: Since 2 does not divide evenly into 5, simply bring down the 5.

Step 4: Continue dividing by a prime number until only 1's are left as quotients:

$$\begin{array}{l} 3\overline{)6 - 12 - 15} \\ 2\overline{)2 - 4 - 5} \\ 5\overline{)1 - 2 - 5} \\ 2\overline{)1 - 2 - 1} \\ 1 - 1 - 1 \end{array}$$

Step 5: Multiply all prime number divisors; the product is the LCD:

$$3 \times 2 \times 5 \times 2 = 60 \longleftarrow \text{LCD}$$

Now that we found the LCD of 60, we can finish the problem by raising each fraction to 60ths and adding the numerators:

$$\begin{array}{rl} & \frac{1}{6} = \frac{10}{60} \\ + & \frac{5}{12} = \frac{25}{60} \\ & \frac{4}{15} = \frac{16}{60} \\ \hline & \phantom{\frac{4}{15} =} \frac{51}{60} = \frac{51 \div 3}{60 \div 3} = \frac{17}{20} \end{array}$$

Remember. Fractions with unlike terms must be converted to equivalent fractions before they can be added.

Name ______________________ Date ______________________

UNIT 21 Exercises

Add these fractions. Change improper fractions to whole numbers or mixed numbers and reduce all final answers to lowest terms.—*Hint:* If both denominators are prime, just multiply them together to get the least common denominator.

ANSWERS

1. ____________
2. ____________
3. ____________
4. ____________
5. ____________
6. ____________

1. $\frac{3}{4} + \frac{1}{6}$

2. $\frac{5}{7} + \frac{2}{9}$

3. $\frac{7}{8} + \frac{3}{4}$

4. $\frac{2}{5} + \frac{1}{3}$

5. $\frac{9}{11} + \frac{3}{5}$

6. $\frac{1}{4} + \frac{2}{3}$

7. $\frac{4}{5}$

$+ \frac{5}{7}$

8. $\frac{5}{8}$

$+ \frac{7}{9}$

ANSWERS

7. ____________

8. ____________

9. ____________

10. ____________

11. ____________

12. ____________

9. $\frac{2}{9}$

$+ \frac{3}{4}$

10. $\frac{7}{8}$

$+ \frac{3}{5}$

11. $\frac{6}{7}$

$+ \frac{3}{4}$

12. $\frac{8}{13}$

$+ \frac{5}{26}$

1. $\frac{3}{4} + \frac{1}{6}$

2. $\frac{5}{7} + \frac{2}{9}$

3. $\frac{7}{8} + \frac{3}{4}$

4. $\frac{2}{5} + \frac{1}{3}$

6. $\frac{1}{4} + \frac{2}{3}$

7. $\frac{4}{5}$
$+ \frac{5}{7}$

8. $\frac{5}{8}$
$+ \frac{7}{9}$

ANSWERS

7. ____________
8. ____________
9. ____________
10. ____________
11. ____________
12. ____________

9. $\frac{2}{9}$
$+ \frac{3}{4}$

10. $\frac{7}{8}$
$+ \frac{3}{5}$

11. $\frac{6}{7}$
$+ \frac{3}{4}$

12. $\frac{8}{13}$
$+ \frac{5}{26}$

Name ____________________ Date ____________________

13. $\frac{2}{15} + \frac{7}{30}$

14. $\frac{13}{15} + \frac{11}{45}$

ANSWERS

13. ______

14. ______

15. ______

16. ______

17. ______

15. $\frac{7}{25} + \frac{12}{30}$

16. $\frac{5}{6} + \frac{2}{3} + \frac{3}{4}$

17. $\frac{7}{8} + \frac{3}{8} + \frac{2}{9}$

18. $$\begin{array}{r} \frac{12}{17} \\ \frac{12}{17} \\ + \; \frac{11}{34} \\ \hline \end{array}$$

19. $$\begin{array}{r} \frac{21}{25} \\ \frac{8}{15} \\ + \; \frac{13}{20} \\ \hline \end{array}$$

ANSWERS

18. ____________

19. ____________

20. ____________

20. $$\begin{array}{r} \frac{3}{7} \\ \frac{2}{21} \\ + \; \frac{11}{42} \\ \hline \end{array}$$

Name ______________________ Date ______________________

UNIT 21 Applications

1. On a trip to the local health food store, Bill bought $\frac{1}{2}$ pound of sunflower seeds, $\frac{1}{4}$ pound of pumpkin seeds, $\frac{3}{4}$ pound of oat bran, $\frac{3}{4}$ pound of wheat germ, and $\frac{7}{8}$ pound of brown rice. What was the total weight of Bill's purchase?

ANSWERS

1. ____________
2. ____________
3. ____________
4. ____________

2. Dennis jogs $\frac{3}{4}$ mile every morning and $\frac{1}{2}$ mile every evening. How many miles does Dennis jog each day?

3. A cake recipe calls for $\frac{3}{4}$ cup of sugar for the cake and $\frac{1}{6}$ cup of sugar for the frosting. Altogether, how much sugar is needed?

4. Jill spends $\frac{1}{4}$ of her paycheck on food, $\frac{1}{3}$ on rent, $\frac{1}{6}$ on entertainment, and $\frac{1}{5}$ on miscellaneous expenses. What total fraction of her pay does Jill spend?

5. On Monday, Nicholas studied business math for $\frac{3}{4}$ hour, English for $\frac{3}{4}$ hour, and history for $\frac{2}{3}$ hour. How many hours did John spend studying?

ANSWERS

5. __________

6. __________

7. __________

6. Debbie bought $\frac{1}{8}$ pound of hard candy, $\frac{3}{5}$ pound of chocolate candy, and $\frac{1}{6}$ pound of nut-filled candy. How much candy did she buy altogether?

7. To make a bookcase, Bob cut these pieces from a long board: $\frac{2}{3}$ foot, $\frac{1}{8}$ foot, and $\frac{1}{6}$ foot. How many feet of board were cut?

Name ______________________ Date ______________________

8. At the market, Elvira bought $\frac{8}{9}$ pound of tomatoes, $\frac{1}{3}$ pound of squash, and $\frac{5}{6}$ pound of potatoes. How many pounds of vegetables did Elvira buy?

ANSWERS

8. __________

9. __________

10. __________

9. Ted hand-waxed three cars. He used $\frac{3}{5}$ can of wax on the first car, $\frac{2}{3}$ can of wax on the second car, and $\frac{3}{8}$ can of wax on the third car. How many cans of wax did Ted use on all three cars?

10. Jillian and Charlie bought $\frac{7}{8}$ acre of land on June 1. On June 19, they bought another $\frac{1}{2}$ acre. How much land did they buy altogether?

Work Space

UNIT 22

Adding Mixed Numbers

As you will remember, a mixed number is a whole number and a fraction. Thus, when adding mixed numbers, you should work in separate steps, as follows.

To add mixed numbers:

Step 1: Add the fractions.

Step 2: Add the whole numbers.

Step 3: Combine the totals from Steps 1 and 2.

EXAMPLE: Add $10\frac{1}{2} + 8\frac{2}{5}$

Solution:

$$\begin{array}{rrcr} & 10\frac{1}{2} & & 10\frac{5}{10} \\ + & 8\frac{2}{5} & = & 8\frac{4}{10} \\ \hline & & & 18\frac{9}{10} \end{array}$$

Step 1. Add the fractions. The LCD is 10. Raise each denominator to 10ths and add the numerators: $\frac{5}{10} + \frac{4}{10} = \frac{9}{10}$.

Step 3. Combine the totals: $18 + \frac{9}{10} = 18\frac{9}{10}$.

Step 2. Add the whole numbers: 10 + 8 = 18.

When adding mixed numbers, the sum of the fractions may be an improper fraction. If this happens, convert the improper fraction to a mixed number, then add the whole numbers.

EXAMPLE: Add: $12\frac{2}{3} + 8\frac{3}{4}$

Solution:

Step 2: 12 + 8 = 20

Step 1: $\frac{8}{12} + \frac{9}{12} = \frac{17}{12}$

$$\begin{array}{rrcr} & 12\frac{2}{3} & & 12\frac{8}{12} \\ + & 8\frac{3}{4} & = & 8\frac{9}{12} \\ \hline & & & 20\frac{17}{12} \end{array}$$

$\frac{17}{12} = 1\frac{5}{12}$

Step 3: $20 + 1\frac{5}{12} = 21\frac{5}{12}$

Answer $= 21\frac{5}{12}$

Note that the improper fraction $\frac{17}{12}$ was changed to a mixed number, $1\frac{5}{12}$; the mixed number was then added to the total of the whole numbers: $20 + 1\frac{5}{12} = 21\frac{5}{12}$.

Work Space

Name ______________________ Date ______________________

UNIT 22 Exercises

Add these mixed numbers. Reduce final answers to lowest terms.

1. $8\frac{3}{4}$
   $\underline{7\frac{1}{2}}$

2. $5\frac{7}{8}$
   $\underline{4\frac{4}{5}}$

3. $8\frac{1}{3}$
   $\underline{3\frac{2}{7}}$

4. $9\frac{3}{5}$
   $\underline{3\frac{1}{6}}$

5. $3\frac{4}{7}$
   $\underline{2\frac{3}{8}}$

6. $7\frac{1}{4}$
   $\underline{5\frac{2}{9}}$

7. $9\frac{3}{8}$
   $\underline{7\frac{5}{7}}$

8. $3\frac{2}{5}$
   $\underline{2\frac{1}{9}}$

9. $15\frac{2}{9}$
   $5\frac{3}{9}$
   $\underline{17\frac{4}{5}}$

ANSWERS

1. ______
2. ______
3. ______
4. ______
5. ______
6. ______
7. ______
8. ______
9. ______

10. $6\frac{7}{8}$

$25\frac{1}{5}$

$\underline{24\frac{3}{4}}$

11. $23\frac{5}{9}$

$37\frac{8}{9}$

$\underline{55\frac{7}{8}}$

ANSWERS

10. ____________

11. ____________

12. ____________

13. ____________

14. ____________

15. ____________

12. $15\frac{14}{15}$

$14\frac{12}{25}$

$\underline{21\frac{13}{50}}$

13. $22\frac{11}{14}$

$15\frac{9}{56}$

$\underline{7\frac{15}{28}}$

14. $19\frac{11}{16}$

$9\frac{3}{4}$

$\underline{17\frac{13}{48}}$

15. $18\frac{13}{30}$

$19\frac{4}{5}$

$\underline{28\frac{7}{10}}$

Name ______________________ Date ______________________

16. $14\frac{5}{7}$

$18\frac{2}{3}$

$\underline{12\frac{9}{14}}$

ANSWERS

16. ______________

Work Space

Name ________________________________ Date ________________________________

UNIT 22 Applications

ANSWERS

1. ____________
2. ____________
3. ____________

1. John worked these hours last week:

Monday	—	$8\frac{1}{2}$ hours
Tuesday	—	$9\frac{1}{2}$ hours
Wednesday	—	$10\frac{1}{4}$ hours
Thursday	—	$8\frac{2}{3}$ hours
Friday	—	$8\frac{1}{2}$ hours

How many total hours did John work during the week?

2. On Tuesday, Joanie sold the following pieces of material: $2\frac{1}{2}$ feet of linen, $3\frac{1}{3}$ feet of silk, and $4\frac{1}{8}$ feet of velvet. How many total feet of material did she sell?

3. To build a dog house, Tom bought $2\frac{1}{2}$ dozen nails and $3\frac{3}{4}$ dozen tacks. How many dozen nails and tacks did Tom buy?

4. A local restaurant bought $4\frac{1}{2}$ pounds of apples, $3\frac{1}{3}$ pounds of tomatoes, $12\frac{3}{4}$ pounds of potatoes, and $1\frac{1}{2}$ pounds of cherries. How many pounds of fruit and vegetables did the restaurant buy?

ANSWERS

4. ____________

5. ____________

6. ____________

5. Julio's jogging schedule for the week is: Monday, $5\frac{1}{2}$ miles; Tuesday, $6\frac{1}{4}$ miles; Wednesday, $6\frac{3}{8}$ miles; Thursday, $8\frac{1}{4}$ miles; Friday, $8\frac{1}{2}$ miles. How many miles will Julio jog in the week?

6. Judy needs $3\frac{5}{8}$ yards of cotton fabric and $4\frac{2}{3}$ yards of linen to make a dress. How much material does she need?

1. ____________
2. ____________
3. ____________

Name ______________________ Date ______________________

7. The following pieces were cut from a pipe: $5\frac{1}{8}$ feet, $6\frac{5}{7}$ feet, and $4\frac{7}{9}$ feet. How many total feet were cut from the pipe?

ANSWERS

7. ____________

8. ____________

9. ____________

8. In a pie eating contest, Sam ate $8\frac{3}{4}$ pies, Bill ate $6\frac{1}{5}$ pies, and Todd ate $7\frac{1}{6}$ pies. How many pies were eaten during the contest?

9. On a summer vacation, the Kellner family drove the following distances to see various attractions: $25\frac{1}{8}$ miles, $125\frac{6}{7}$ miles, $345\frac{3}{4}$ miles, and $211\frac{1}{4}$ miles. How many total miles did they drive?

10. On a spring shape-up program, Marcus lost $10\frac{3}{8}$ pounds and Elaine lost $7\frac{2}{3}$ pounds. How much total weight did the couple lose?

ANSWERS

10. ____________

UNIT 23

Subtracting Fractions

The process of subtracting fractions is very similar to the process of adding fractions. To subtract fractions, follow these steps:

Step 1: Line up the fractions vertically, the smaller under the larger.

Step 2: Find the least common denominator, and raise the fractions to equivalent terms.

Step 3: Subtract the numerators.

Step 4: Place the difference between the numerators over common denominator.

Step 5: Reduce the final answer to lowest terms, if needed.

EXAMPLE 1: Subtract: $\frac{7}{8} - \frac{3}{4}$

Solution:

We can observe that the LCD is 8. Raise the fractions to 8ths and subtract the numerators:

$$\begin{array}{r} \frac{7}{8} = \frac{7}{8} \\ -\frac{3}{4} = \frac{6}{8} \\ \hline \frac{1}{8} \end{array}$$

$8 \div 4 = 2; 2 \times 3 = 6$

$\frac{7}{8} - \frac{6}{8} = \frac{1}{8}$

EXAMPLE 2: Subtract: $\frac{3}{10} - \frac{2}{15}$

Solution:

The LCD is 30. Raise the fractions to 30ths and subtract the numerators:

$$\begin{array}{r} \frac{3}{10} = \frac{9}{30} \\ -\frac{2}{15} = \frac{4}{30} \\ \hline \frac{5}{30} \end{array} = \frac{5 \div 5}{30 \div 5} = \frac{1}{6}$$

Divide both terms by 5 to reduce the answer to lowest terms.

Remember. When subtracting fractions, only subtract the numerators.

Work Space

Name ______________________ Date ______________________

UNIT 23 Exercises

Subtract these fractions. Reduce answers to lowest terms.

1. $\frac{3}{4} - \frac{1}{2}$
2. $\frac{4}{5} - \frac{3}{4}$
3. $\frac{7}{8} - \frac{5}{7}$
4. $\frac{8}{9} - \frac{2}{3}$
5. $\frac{6}{7} - \frac{3}{5}$
6. $\frac{4}{7} - \frac{3}{8}$
7. $\frac{2}{3} - \frac{1}{6}$
8. $\frac{5}{6} - \frac{4}{9}$
9. $\frac{3}{8} - \frac{2}{9}$
10. $\frac{4}{5} - \frac{5}{8}$
11. $\frac{1}{3} - \frac{1}{4}$
12. $\frac{8}{9} - \frac{5}{8}$

ANSWERS

1. ______
2. ______
3. ______
4. ______
5. ______
6. ______
7. ______
8. ______
9. ______
10. ______
11. ______
12. ______

13. $\frac{4}{5} - \frac{2}{7}$

14. $\frac{6}{7} - \frac{4}{9}$

15. $\frac{5}{6} - \frac{7}{9}$

16. $\frac{5}{9} - \frac{3}{8}$

17. $\frac{3}{7} - \frac{2}{9}$

18. $\frac{1}{2} - \frac{4}{9}$

19. $\frac{2}{3} - \frac{7}{26}$

20. $\frac{11}{12} - \frac{11}{20}$

21. $\frac{8}{9} - \frac{1}{3}$

22. $\frac{3}{4} - \frac{2}{9}$

23. $\frac{1}{4} - \frac{2}{25}$

24. $\frac{3}{5} - \frac{2}{15}$

25. $\frac{10}{11} - \frac{2}{3}$

ANSWERS

13. ____________

14. ____________

15. ____________

16. ____________

17. ____________

18. ____________

19. ____________

20. ____________

21. ____________

22. ____________

23. ____________

24. ____________

25. ____________

Name ______________________ Date ______________________

UNIT 23 Applications

ANSWERS

1. ________
2. ________
3. ________
4. ________

1. Gerri took $\frac{3}{4}$ pound of grapes to work with her. At lunch she ate $\frac{1}{6}$ pound. How much was left?

2. Which is larger, $\frac{7}{8}$ or $\frac{7}{9}$? How much larger?

3. Jeffrey ate $\frac{3}{4}$ of a pizza and Clinton ate $\frac{2}{3}$ of a pizza. How much more did Jeffrey eat?

4. A cake recipe calls for $\frac{3}{4}$ cup of white sugar and $\frac{1}{8}$ cup of brown sugar. How much more white sugar is needed than brown sugar?

5. Roger jogged $\frac{9}{10}$ of a mile in 6 minutes and Alice jogged $\frac{7}{9}$ of a mile in 6 minutes. Who is faster? How much farther does the faster runner go in 6 minutes?

ANSWERS

5. ____________

6. ____________

7. ____________

6. Lang ordered a quarter pound burger ($\frac{1}{4}$ pound) for lunch. He ate about one-fifth ($\frac{1}{5}$) pound. How much was left?

7. The Kingrey family bought $\frac{3}{4}$ pound of pecan candy and ate $\frac{1}{2}$ pound. How much was left?

Name ______________________ Date ______________________

8. Kathy bought $\frac{9}{10}$ yard of material to make a skirt. If she used $\frac{7}{9}$ yard, how much was left?

ANSWERS

8. ____________

9. ____________

10. ____________

9. On June 10, Jeff bought $\frac{7}{8}$ ton of grain. On July 3, Jeff sold $\frac{5}{6}$ ton of the grain. How much did he have left?

10. When Ellen started her trip, she had a full tank of gas ($\frac{1}{1}$). At the end of the trip, she only had $\frac{1}{8}$ tank. How much gas did she use?

Work Space

UNIT
24
Subtracting Mixed Numbers

Subtracting mixed numbers is very similar to adding mixed numbers. To subtract mixed numbers, you follow these steps:

Step 1: Line up the numbers vertically, the smaller under the larger.

Step 2: Find a common denominator and subtract the fractions.

Step 3: Subtract the whole numbers.

Step 4: Reduce the final answer to lowest terms, if needed.

EXAMPLE: Subtract $5\frac{3}{4} - 3\frac{1}{3}$
Solution:

The LCD is 12; raise the fractions to equivalent fractions with a common denominator of 12.

$$\begin{array}{rcl} 5\frac{3}{4} & = & 5\frac{9}{12} \\ -\ 3\frac{1}{3} & = & 3\frac{4}{12} \\ \hline & & 2\frac{5}{12} \end{array}$$

$12 \div 4 = 3;\ 3 \times 3 = 9$

$12 \div 3 = 4;\ 4 \times 1 = 4$

$\frac{9}{12} - \frac{4}{12} = \frac{5}{12}$

$5 - 3 = 2$

SUBTRACTING MIXED NUMBERS WHEN BORROWING IS NEEDED

When subtracting mixed numbers, the fraction in the bottom number (the subtrahend) is sometimes larger than the fraction in the top number (the minuend). When this happens, you borrow 1 from the top whole number and add it to the top fraction. Let's look at a couple of examples.

EXAMPLE 1: Subtract: $8\frac{1}{3} - 5\frac{3}{5}$
Solution:

The LCD is 15; raise the fractions to equivalent fractions with a common denominator of 15.

$$\begin{array}{rcl} 8\frac{1}{3} & = & 8\frac{5}{15} \\ -\ 5\frac{3}{5} & = & 5\frac{9}{15} \\ \hline \end{array}$$

$15 \div 3 = 5;\ 5 \times 1 = 5$

$15 \div 5 = 3;\ 3 \times 3 = 9$

Notice that $\frac{9}{15}$ is larger than $\frac{5}{15}$. Since we cannot subtract $\frac{9}{15}$ from $\frac{5}{15}$, we have to borrow a 1 from the 8; change the 8 to 7 and write the 1 as $\frac{15}{15}$, and add it to $\frac{5}{15}$: $\frac{15}{15} + \frac{5}{15} = \frac{20}{15}$. Then, rewrite the problem to show the borrowing:

$$\begin{array}{r} \overset{7}{\not{8}}\frac{5}{15} + \frac{15}{15} = 7\frac{20}{15} \\ -\ 5\frac{9}{15} \qquad = 5\frac{9}{15} \\ \hline 2\frac{11}{15} \end{array}$$

$7 - 5 = 2$

$\frac{20}{15} - \frac{9}{15} = \frac{11}{15}$

EXAMPLE 2: Subtract: $12\frac{3}{4} - 6\frac{7}{8}$
Solution:

The LCD is 8.

$$\begin{array}{r} 12\frac{3}{4} = 12\frac{6}{8} \\ -\ \ 6\frac{7}{8} = \ \ 6\frac{7}{8} \\ \hline \end{array}$$

$8 \div 4 = 2; 2 \times 3 = 6$

Since $\frac{7}{8}$ is larger than $\frac{6}{8}$, we must borrow a 1 from the 12. Change the 12 to 11 and write the 1 as $\frac{8}{8}$, and add it to $\frac{6}{8}$: $\frac{8}{8} + \frac{6}{8} = \frac{14}{8}$. Rewrite the problem to show the borrowing.

$$\begin{array}{r} \overset{11}{\not{12}}\frac{6}{8} + \frac{8}{8} = 11\frac{14}{8} \\ 6\frac{7}{8} \qquad = 6\frac{7}{8} \\ \hline 5\frac{7}{8} \end{array}$$

$11 - 6 = 5$

$\frac{14}{8} - \frac{7}{8} = \frac{7}{8}$

SUBTRACTING A MIXED NUMBER FROM A WHOLE NUMBER

To subtract a mixed number from a whole number, follow the steps in the next example.

EXAMPLE: Subtract: $18 - 12\frac{1}{8}$
Solution:

Step 1: Write the problem vertically:

$$\begin{array}{r} 18 \\ -\ 12\frac{1}{8} \\ \hline \end{array}$$

Step 2: Borrow a 1 from the whole number and convert the 1 to a fraction equivalent to the fraction in the bottom number; then subtract.

$$\begin{array}{r} 18 \ = 17\frac{8}{8} \\ -\ 12\frac{1}{8} = 12\frac{1}{8} \\ \hline 5\frac{7}{8} \end{array}$$

Remember. Anytime a number is written over itself, the value is always equal to 1. So, $\frac{12}{12}$, $\frac{7}{7}$, $\frac{55}{55}$, $\frac{500}{500}$, as well as any other number written over itself, are equal to a value of 1.

Name ______________________________ Date ______________________________

UNIT 24 Exercises

Subtract these mixed numbers. Reduce answers to lowest terms.

1. $15\frac{3}{5} - 8\frac{1}{3}$
2. $18\frac{5}{6} - 7\frac{3}{4}$
3. $12\frac{4}{7} - 10\frac{1}{2}$
4. $25\frac{3}{8} - 12\frac{2}{3}$
5. $13\frac{7}{8} - 12\frac{4}{7}$
6. $34\frac{5}{6} - 31\frac{3}{5}$
7. $77\frac{3}{7} - 45\frac{4}{9}$
8. $11\frac{2}{5} - 10\frac{1}{3}$
9. $24\frac{4}{5} - 16\frac{6}{7}$

ANSWERS

1. ____________
2. ____________
3. ____________
4. ____________
5. ____________
6. ____________
7. ____________
8. ____________
9. ____________

10. $39\frac{6}{7}$
$\underline{21\frac{2}{3}}$

11. $45\frac{1}{8}$
$\underline{32\frac{5}{7}}$

12. $15\frac{7}{9}$
$\underline{12\frac{8}{9}}$

ANSWERS

10. ______________
11. ______________
12. ______________
13. ______________
14. ______________
15. ______________
16. ______________
17. ______________
18. ______________

13. $31\frac{11}{12}$
$\underline{20\frac{9}{18}}$

14. $76\frac{12}{13}$
$\underline{35\frac{7}{39}}$

15. $18\frac{17}{40}$
$\underline{12\frac{15}{30}}$

16. $27\frac{21}{40}$
$\underline{24\frac{17}{20}}$

17. 15
$\underline{8\frac{3}{5}}$

18. 25
$\underline{12\frac{7}{8}}$

Name ________________________ Date ________________________

19. $\begin{array}{r} 21 \\ \underline{15\frac{5}{6}} \end{array}$

20. $\begin{array}{r} 19 \\ \underline{10\frac{10}{11}} \end{array}$

21. $21\frac{7}{12} - 14\frac{2}{3}$

ANSWERS

19. ____________
20. ____________
21. ____________
22. ____________
23. ____________
24. ____________
25. ____________
26. ____________
27. ____________

22. $35\frac{1}{5} - 7\frac{3}{4}$

23. $45\frac{1}{8} - 17\frac{3}{7}$

24. $14\frac{2}{3} - 7\frac{5}{9}$

25. $18\frac{4}{11} - 9\frac{4}{5}$

26. $44\frac{5}{12} - 12\frac{11}{12}$

27. $15 - 12\frac{1}{8}$

28. $25 - 15\frac{2}{9}$

29. $45 - 4\frac{5}{6}$

30. $10 - \frac{1}{2}$

ANSWERS

28. ____________

29. ____________

30. ____________

Name ______________________ Date ______________________

UNIT 24 Applications

1. Bill McKay owned $10\frac{1}{2}$ acres of land. At his daughter's marriage, Bill gave her $2\frac{1}{4}$ acres as a home site. How much land did Bill have left?

ANSWERS

1. ____________
2. ____________
3. ____________

2. Before cooking, a large steak weighed $4\frac{1}{8}$ pounds. After cooking, it weighed $3\frac{4}{5}$ pounds. How much weight was lost in cooking?

3. Lynn bought a bag of flour that weighed 48 ounces. She then used $12\frac{3}{4}$ ounces to bake a cake. How much was left?

4. How many feet would be left if $10\frac{1}{2}$ feet are cut from a board that is $18\frac{1}{8}$ feet long?

ANSWERS

4. ____________
5. ____________
6. ____________
7. ____________

5. John bought $12\frac{1}{4}$ acres of land on June 1. He bought another $25\frac{3}{8}$ acres on July 15. He then sold $10\frac{1}{2}$ acres on August 31. How many acres of land did John have left?

6. Jill's car has a $15\frac{1}{2}$ gallon gas tank. She believes that it will take $10\frac{1}{4}$ gallons to drive to a friend's home, who lives in another state. How much gas should Jill have left?

7. The stock of ABC Corporation opened on Tuesday morning at $48\frac{1}{2}$ ($48.50) a share. It closed Tuesday evening at $48\frac{1}{8}$. How much value did the stock lose?

Name ______________________ Date ______________________

8. A piece of material is $35\frac{7}{8}$ yards long. If $12\frac{2}{5}$ yards are sold, how much material is left?

ANSWERS

8. ____________

9. ____________

10. ____________

9. Pam bought $3\frac{1}{2}$ dozen oranges. If $2\frac{1}{4}$ dozen were used to make juice, how many dozen remain?

10. Scott bought 15 gallons of paint to paint his house. When the job was finished, Scott had $3\frac{1}{3}$ gallons left. How much paint did he use?

Work Space

UNIT 25 Multiplying Fractions

To multiply fractions, you multiply like terms: numerators times numerators, and denominators times denominators. Answers should be reduced to lowest terms, if needed.

EXAMPLE 1: Multiply: $\frac{4}{5} \times \frac{3}{4}$
Solution:

Step 1: Multiply like terms of the fractions:

$$\frac{4}{5} \times \frac{3}{4} = \frac{12}{20}$$

Step 2: Reduce to lowest terms:

$$\frac{12}{20} = \frac{12 \div 4}{20 \div 4} = \frac{3}{5}$$

EXAMPLE 2: Multiply: $\frac{2}{3} \times \frac{1}{2} \times \frac{3}{4}$
Solution:

Step 1: Multiply like terms:

$$\frac{2}{3} \times \frac{1}{2} \times \frac{3}{4} = \frac{6}{24}$$

Step 2: Reduce to lowest terms:

$$\frac{6}{24} = \frac{6 \div 6}{24 \div 6} = \frac{1}{4}$$

CANCELLATION

You can simplify the multiplication of many fractions by using **cancellation.** Cancellation can be used when a numerator *and* a denominator can be evenly divided by the same whole number.

EXAMPLE 1: Multiply using cancellation: $\frac{3}{4} \times \frac{7}{9}$
Solution:

$$\frac{\overset{1}{\cancel{3}}}{4} \times \frac{7}{\underset{3}{\cancel{9}}} = \frac{7}{12}$$

The 3 and the 9 are both evenly divisible by 3: 3 ÷ 3 = 1; 9 ÷ 3 = 3.

EXAMPLE 2: Multiply using cancellation: $\frac{4}{5} \times \frac{7}{10}$
Solution:

$$\frac{\overset{2}{\cancel{4}}}{5} \times \frac{7}{\underset{5}{\cancel{10}}} = \frac{14}{25}$$

The 4 and the 10 are both evenly divisible by 2: 4 ÷ 2 = 2; 10 ÷ 2 = 5.

It should be stressed that cancellation can be used only when a numerator and a denominator are evenly divisible by the same whole number. The numerators and denominators do not have to be in the same fraction, and you can cancel more than once in a problem.

Cancellation is especially helpful when you are multiplying more than two fractions. Let's look at an example.

EXAMPLE: Multiply $\frac{7}{8} \times \frac{3}{4} \times \frac{16}{21} \times \frac{2}{9}$

Solution:

$$\frac{\cancel{7}^{1}}{\cancel{8}_{1}} \times \frac{\cancel{3}^{1}}{\cancel{4}_{\cancel{2}_{1}}} \times \frac{\cancel{16}^{\cancel{2}^{1}}}{\cancel{21}_{\cancel{3}_{1}}} \times \frac{\cancel{2}^{1}}{9} = \frac{1}{9}$$

We made quite a few cancellations in this problem. Here is the order in which they were made:

1. 7 and 21: 7 ÷ 7 = 1; 21 ÷ 7 = 3.
2. 16 and 8: 8 ÷ 8 = 1; 16 ÷ 8 = 2.
3. 3 and 3: 3 ÷ 3 = 1; 3 ÷ 3 = 1.
4. 4 and 2: 4 ÷ 2 = 2; 2 ÷ 2 = 1.
5. 2 and 2: 2 ÷ 2 = 1; 2 ÷ 2 = 1.

HELPFUL HINT

Sometimes you will have to work with problems such as "what is $\frac{1}{2}$ of $\frac{3}{5}$?" When a problem is given like this, the word "of" means to multiply. So, $\frac{1}{2}$ of $\frac{3}{5}$ is:

$$\frac{1}{2} \times \frac{3}{5} = \frac{3}{10}$$

Remember.

1. When multiplying fractions, multiply only like terms.
2. Use cancellation only when there is a whole number that can be divided evenly into **both** a numerator and a denominator; **you never cancel using just numerators or just denominators.**

Name ______________________ Date ______________________

UNIT 25 Exercises

Multiply each of these problems. Use cancellation where possible.

1. $\frac{3}{4} \times \frac{4}{5}$ 2. $\frac{7}{8} \times \frac{5}{9}$ 3. $\frac{8}{9} \times \frac{2}{7}$ 4. $\frac{3}{8} \times \frac{4}{5}$ 5. $\frac{5}{6} \times \frac{7}{8}$

6. $\frac{1}{5} \times \frac{4}{7}$ 7. $\frac{3}{5} \times \frac{4}{9}$ 8. $\frac{2}{3} \times \frac{6}{9}$ 9. $\frac{1}{4} \times \frac{5}{7}$ 10. $\frac{2}{7} \times \frac{7}{9}$

11. $\frac{8}{9} \times \frac{3}{4}$ 12. $\frac{3}{5} \times \frac{5}{9}$ 13. $\frac{5}{7} \times \frac{7}{8}$ 14. $\frac{3}{5} \times \frac{7}{9}$ 15. $\frac{1}{3} \times \frac{4}{5}$

ANSWERS

1. ____________
2. ____________
3. ____________
4. ____________
5. ____________
6. ____________
7. ____________
8. ____________
9. ____________
10. ____________
11. ____________
12. ____________
13. ____________
14. ____________
15. ____________

16. $\frac{4}{5} \times \frac{3}{8}$ 17. $\frac{2}{5} \times \frac{5}{8}$ 18. $\frac{8}{9} \times \frac{7}{8}$ 19. $\frac{2}{7} \times \frac{7}{8}$ 20. $\frac{1}{5} \times \frac{5}{6}$

21. $\frac{7}{9} \times \frac{3}{5}$ 22. $\frac{2}{5} \times \frac{1}{2} \times \frac{3}{4}$ 23. $\frac{3}{4} \times \frac{4}{7} \times \frac{7}{9}$ 24. $\frac{4}{5} \times \frac{6}{8} \times \frac{8}{9}$

25. $\frac{4}{7} \times \frac{7}{8} \times \frac{2}{3}$ 26. $\frac{2}{5} \times \frac{5}{8} \times \frac{8}{9}$ 27. $\frac{2}{6} \times \frac{6}{8} \times \frac{3}{4}$ 28. $\frac{3}{7} \times \frac{7}{9} \times \frac{3}{4}$

29. $\frac{2}{5} \times \frac{4}{7} \times \frac{6}{9} \times \frac{1}{8}$ 30. $\frac{2}{3} \times \frac{5}{6} \times \frac{8}{9} \times \frac{1}{2}$ 31. $\frac{3}{7} \times \frac{7}{9} \times \frac{5}{6} \times \frac{3}{5}$

ANSWERS

16. ______________
17. ______________
18. ______________
19. ______________
20. ______________
21. ______________
22. ______________
23. ______________
24. ______________
25. ______________
26. ______________
27. ______________
28. ______________
29. ______________
30. ______________
31. ______________

Name ______________________ Date ______________________

32. $\frac{3}{5} \times \frac{5}{7} \times \frac{7}{8} \times \frac{3}{4} \times \frac{8}{9}$

33. $\frac{12}{17} \times \frac{34}{48}$

34. $\frac{3}{4} \times \frac{7}{11} \times \frac{11}{12} \times \frac{13}{15}$

ANSWERS

32. ____________

33. ____________

34. ____________

Work Space

Name ______________________ Date ______________________

UNIT 25 Applications

1. On Tuesday, the Great Southern Mining Company extracted $\frac{1}{2}$ ton of iron ore. If $\frac{1}{8}$ of this is actually good enough to be used to make iron, how much is usable?

ANSWERS

1. ____________
2. ____________
3. ____________
4. ____________

2. Lisa Day bought $\frac{3}{4}$ dozen eggs and used $\frac{1}{3}$ of them. How many eggs did she use?

3. What is $\frac{3}{4}$ of $\frac{5}{20}$?

4. Find $\frac{1}{2}$ of $\frac{2}{3}$.

5. Nelson Holmes ran $\frac{1}{2}$ of a $\frac{3}{4}$-mile trail. How far did he run?

ANSWERS

5. ____________

6. ____________

7. ____________

6. The Epson Grocery Company spends $\frac{3}{5}$ of its budget on salaries. If $\frac{2}{3}$ of this goes for executive salaries, what part of its total budget goes for other salaries?

7. Which is larger: $\frac{1}{4}$ of $\frac{5}{6}$ or $\frac{1}{5}$ of $\frac{7}{8}$? How much larger?

UNIT
26

Multiplying Mixed Numbers

MULTIPLYING A MIXED NUMBER BY A MIXED NUMBER

To multiply a mixed number by a mixed number, follow these steps:

Step 1: Change the mixed numbers to improper fractions.

Step 2: Cancel, where possible, and multiply across.

Step 3: Change answers that are improper fractions to whole numbers or to mixed numbers.

EXAMPLE: Multiply $2\frac{1}{3} \times 5\frac{1}{4}$

Solution:

Step 1: Change the mixed numbers to improper fractions:

$$2\frac{1}{3} = \frac{(3 \times 2 = 6) + 1}{3} = \frac{7}{3}$$

$$5\frac{1}{4} = \frac{(4 \times 5 = 20) + 1}{4} = \frac{21}{4}$$

Step 2: Cancel and multiply across:

$$\frac{7}{\cancel{3}_{1}} \times \frac{\cancel{21}^{7}}{4} = \frac{49}{4}$$

Step 3: Change improper fraction to a mixed number:

$$\frac{49}{4} = 4\overline{)49} = 12\ \text{R}\ 1 = 12\frac{1}{4}$$

$$\begin{array}{r} 12 \\ 4\overline{)49} \\ \underline{48} \\ 1 \end{array}$$

MULTIPLYING A WHOLE NUMBER BY A MIXED NUMBER

To multiply a whole number by a mixed number, you change the mixed number to an improper fraction, and you change the whole number to an improper fraction; you then cancel, where possible, and multiply across.

EXAMPLE 1: Multiply: $5 \times 3\frac{1}{5}$

Solution:

Step 1: Change the whole number and the mixed number to improper fractions:

$$5 \longrightarrow \frac{5}{1}$$

A whole number is changed to an improper fraction by writing the whole number over the number 1.

$$3\frac{1}{5} = \frac{(5 \times 3 = 15) + 1}{5} = \frac{16}{5}$$

Cancel and multiply across:

$$\frac{\overset{1}{\cancel{5}}}{1} \times \frac{16}{\underset{1}{\cancel{5}}} = \frac{16}{1} = 16$$

EXAMPLE 2: Multiply: $24 \times 6\frac{1}{4}$

Solution:

Step 1: Change the whole number and the mixed number to improper fractions:

$$24 \qquad \frac{24}{1}$$

$$6\frac{1}{4} = \frac{(4 \times 6 = 24) + 1}{4} = \frac{25}{4}$$

Step 2: Cancel and multiply across:

$$\frac{\overset{6}{\cancel{24}}}{1} \times \frac{25}{\underset{1}{\cancel{4}}} = \frac{150}{1} = 150$$

Name ______________________ Date ______________________

UNIT 26 Exercises

Multiply each of these problems. Change answers that are improper fractions to whole numbers or to mixed numbers.

ANSWERS

1. ________
2. ________
3. ________
4. ________
5. ________
6. ________
7. ________
8. ________
9. ________

1. $5\frac{1}{3} \times 3\frac{1}{2}$

2. $7\frac{2}{5} \times 3\frac{1}{8}$

3. $6\frac{1}{8} \times 3\frac{2}{7}$

4. $4\frac{2}{5} \times 5\frac{5}{8}$

5. $6\frac{2}{3} \times 7\frac{5}{9}$

6. $8\frac{6}{7} \times 3\frac{5}{8}$

7. $2\frac{4}{7} \times 4\frac{6}{7}$

8. $3\frac{3}{4} \times 2\frac{2}{9}$

9. $1\frac{8}{9} \times 5\frac{1}{3}$

10. $6\frac{3}{4} \times 2\frac{7}{8}$

11. $2\frac{6}{7} \times 7\frac{4}{5}$

12. $3\frac{2}{9} \times 6\frac{5}{29}$

ANSWERS

10. ____________
11. ____________
12. ____________
13. ____________
14. ____________
15. ____________
16. ____________
17. ____________
18. ____________
19. ____________
20. ____________
21. ____________

13. $2\frac{4}{9} \times 5\frac{4}{8}$

14. $3\frac{8}{9} \times 5\frac{1}{7}$

15. $6\frac{2}{3} \times 4\frac{7}{8}$

16. $3\frac{5}{9} \times 3\frac{3}{8}$

17. $5 \times 3\frac{3}{4}$

18. $15 \times 6\frac{3}{5}$

19. $4\frac{5}{6} \times 7$

20. $6 \times 3\frac{2}{5}$

21. $24 \times 4\frac{1}{8}$

Name ______________________ **Date** ______________________

22. $4\frac{1}{2} \times 12$

ANSWERS

22. __________

Work Space

Name ______________________ Date ______________________

UNIT 26 Applications

1. Joe bought 12 boards that are each $5\frac{1}{2}$ feet long. What is the combined length of the 12 boards?

ANSWERS

1. ____________
2. ____________
3. ____________

2. If $2\frac{3}{4}$ yards of fabric are needed to make one dress, how many yards of fabric would be needed to make 6 dresses?

3. Ronald King bought $2\frac{1}{2}$ dozen oranges for each of his 18 students. How many oranges did Ronald buy altogether?

4. How much would $4\frac{1}{2}$ pounds of steak cost if 1 pound costs $4?

ANSWERS

4. ____________
5. ____________
6. ____________
7. ____________
8. ____________

5. Jimmy is paid $1\frac{1}{2}$ times his regular rate of pay if he works overtime. What is his overtime rate if his regular hourly rate is $6 an hour?

6. Len makes $8 an hour. Last week he worked $35\frac{1}{4}$ hours. How much did he make?

7. Sharon saves $\frac{1}{5}$ of her "take-home pay." How much would she save on a payday where she took home $350?

8. If one Whopper Burger weighs $\frac{3}{8}$ pounds, how much would 12 Whopper Burgers weigh?

Name ______________________ Date ______________________

9. If carpet sells for \$8 a square yard, how much would $22\frac{1}{2}$ square yards sell for?

ANSWERS

9. ________

10. ________

10. What is the cost of $12\frac{1}{8}$ pounds of ground beef if 1 pound costs \$2?

Work Space

UNIT 27 Multiplying Large Mixed Numbers

In Unit 26, we multiplied mixed numbers by changing them to improper fractions. Although this method works well for many problems, it can result in improper fractions that are hard to work with if the mixed numbers are large. For example, if you were multiplying $226\frac{3}{8} \times 312\frac{1}{4}$, changing the mixed numbers to improper fractions would give you:

$$\frac{1{,}811}{8} \times \frac{1{,}249}{4} =$$

As you can see, these improper fractions have large numerators and are somewhat hard to work with. In such situations, it is usually easier to use the **four-step method.** To multiply mixed numbers using the four-step method, follow these steps:

Step 1: Multiply the two fractions.

Step 2: Multiply each whole number by the opposite fraction.

Step 3: Multiply the whole numbers.

Step 4: Add the results from the first three steps.

EXAMPLE 1: Using the four-step method, multiply: $226\frac{3}{8} \times 312\frac{1}{4}$

Solution:

$$\begin{array}{r} 312\frac{1}{4} \\ \times\ 226\frac{3}{8} \\ \hline \frac{3}{32} \\ \\ 117 \\ 56\frac{1}{2} \\ \underline{70{,}512} \\ 70{,}685\frac{19}{32} \end{array}$$

Step 1: Multiply the fractions: $\frac{3}{8} \times \frac{1}{4} = \frac{3}{32}$

Step 2: Multiply whole numbers by opposite fraction:
$312 \times \frac{3}{8} = 117$
$226 \times \frac{1}{4} = 56\frac{1}{2}$

Step 3: Multiply whole numbers:
$312 \times 226 = 70{,}512$

Step 4: Add results:

$$\begin{array}{rcl} \frac{3}{32} & = & \frac{3}{32} \\ 117 & & \\ 56\frac{1}{2} & = & \frac{16}{32} \\ \underline{70{,}512} & & \\ 70{,}685 & & \frac{19}{32} \end{array}$$

MULTIPLYING A LARGE WHOLE NUMBER BY A MIXED NUMBER

To multiply a large whole number by a mixed number, follow the steps in the next example.

EXAMPLE: Multiply $160 \times 30\frac{3}{4}$

Solution:

Step 1: Multiply the whole number, 160, by the whole number part of the mixed number, 30:

$$\begin{array}{r} 160 \\ \times \quad 30 \\ \hline 4{,}800 \end{array}$$

Step 2: Multiply the whole number, 160, by the fraction part of the mixed number, $\frac{3}{4}$:

$$\frac{\overset{40}{\cancel{160}}}{1} \times \frac{3}{\underset{1}{\cancel{4}}} = 120$$

Step 3: Add the results from Steps 1 and 2 to get the final product:

$$\begin{array}{r} 4{,}800 \\ + \quad 120 \\ \hline 4{,}920 \end{array}$$

Answer = 4,920

Name ________________________ Date ________________________

► UNIT 27 Exercises

Multiply each of these problems. Use cancellation where possible.

1. $320\frac{1}{2} \times 120\frac{5}{8}$
2. $245\frac{2}{3} \times 270\frac{4}{5}$
3. $420\frac{2}{7} \times 210\frac{5}{6}$
4. $300\frac{7}{8} \times 255\frac{4}{5}$
5. $428\frac{1}{2} \times 218\frac{1}{6}$
6. $485\frac{3}{4} \times 32\frac{4}{5}$
7. $390\frac{1}{8} \times 40\frac{3}{7}$
8. $210\frac{7}{9} \times 810\frac{3}{7}$
9. $120 \times 18\frac{4}{5}$

ANSWERS

1. ________
2. ________
3. ________
4. ________
5. ________
6. ________
7. ________
8. ________
9. ________

10. $345 \times 12\frac{1}{8}$

11. $765 \times 25\frac{1}{4}$

12. $530 \times 31\frac{9}{10}$

ANSWERS

10. ____________
11. ____________
12. ____________
13. ____________
14. ____________
15. ____________
16. ____________
17. ____________
18. ____________
19. ____________
20. ____________
21. ____________

13. $324 \times 80\frac{1}{6}$

14. $450 \times 16\frac{4}{9}$

15. $386 \times 20\frac{1}{8}$

16. $594 \times 22\frac{5}{6}$

17. $550 \times 25\frac{3}{5}$

18. $300 \times 45\frac{3}{4}$

19. $112 \times 21\frac{3}{16}$

20. $340 \times 18\frac{4}{15}$

21. $262 \times 27\frac{1}{8}$

Name ________________________________ Date ________________________________

22. $138 \times 12\frac{5}{6}$

23. $345 \times 21\frac{2}{9}$

24. $565 \times 32\frac{4}{5}$

ANSWERS

22. __________

23. __________

24. __________

25. __________

25. $310 \times 41\frac{6}{7}$

Work Space

Name ______________________ Date ______________________

UNIT 27 Applications

1. The cost of a pound of a certain industrial cleaner is $\$3\frac{5}{8}$. How much would $12\frac{1}{2}$ pounds of the cleaner cost?

ANSWERS

1. ____________
2. ____________
3. ____________

2. If carpet costs $\$12\frac{1}{2}$ a square yard, how much would $25\frac{3}{4}$ square yards cost?

3. If Eddie drives 55 miles per hour, how far would he drive in $10\frac{1}{2}$ hours?

4. If a construction block weighs $8\frac{2}{3}$ pounds, how much would $22\frac{1}{2}$ blocks weigh?

ANSWERS

4. ____________
5. ____________
6. ____________
7. ____________

5. Steve earns \$8 an hour. How much would he earn in a week in which he worked $39\frac{1}{3}$ hours?

6. If a bottle of cola holds $6\frac{1}{2}$ ounces, how many ounces are in 18 bottles?

7. The stock of Superior Corporation is selling for $33\frac{1}{3}$ (\$$33\frac{1}{3}$) per share. How much would 90 shares cost?

Name ______________________ Date ______________________

8. If it takes $40\frac{1}{8}$ gallons of water to fill a small drum, how many gallons would it take to fill $10\frac{3}{4}$ drums?

ANSWERS

8. ____________

9. ____________

10. ____________

9. It takes $3\frac{1}{2}$ yards of fabric to make a certain coat. How many yards of fabric would it take to make 15 coats?

10. What is the cost of $25\frac{1}{2}$ pounds of ground beef if 1 pound costs $\$2\frac{1}{4}$?

Work Space

UNIT 28

Dividing Fractions

When you divide fractions, you set the problem up with the dividend first, and the divisor second, like this:

$$\frac{3}{4} \div \frac{1}{3}$$

Dividend (points to $\frac{3}{4}$)

Divisor (points to $\frac{1}{3}$)

To divide fractions, you **invert** (turn upside down) the divisor and then multiply the terms.

EXAMPLE 1: Divide: $\frac{3}{4} \div \frac{1}{3}$

Solution:

Step 1: Invert the divisor and change the ÷ symbol to a × symbol:

$$\frac{3}{4} \div \frac{1}{3} \quad\rightarrow\quad \frac{3}{4} \times \frac{3}{1}$$

Don't change the first fraction

Remember: the divisor is the second fraction.

To invert, turn upside down: $\frac{1}{3} = \frac{3}{1}$

Step 2: Multiply across:

$$\frac{3}{4} \times \frac{3}{1} = \frac{9}{4}$$

Step 3: Change the improper fraction, $\frac{9}{4}$, to a mixed number:

$$\frac{9}{4} = 4\overline{)9} \quad 9 - 8 = 1 \quad \text{quotient } 2 = 2\tfrac{1}{4}$$

EXAMPLE 2: Divide: $\frac{7}{8} \div \frac{5}{6}$

Solution:

Step 1: Invert the divisor and change the ÷ symbol to a × symbol:

$$\frac{7}{8} \div \frac{5}{6} \quad\rightarrow\quad \frac{7}{8} \times \frac{6}{5}$$

Step 2: Cancel and multiply across:

$$\frac{7}{\cancel{8}_{4}} \times \frac{\cancel{6}^{3}}{5} = \frac{21}{20}$$

Step 3: Change the improper fraction, $\frac{21}{20}$, to a mixed number:

$$\frac{21}{20} = 20\overline{)21} \quad 21 - 20 = 1 \quad \text{quotient } 1 = 1\tfrac{1}{20}$$

DIVIDING A WHOLE NUMBER BY A FRACTION

To divide a whole number by a fraction, rewrite the whole number as an improper fraction, invert the divisor, and multiply.

EXAMPLE 1: Divide: $6 \div \frac{3}{4}$

Solution:

Step 1: Rewrite the whole number as an improper fraction:

$\frac{6}{1} \div \frac{3}{4}$

Remember: to make a whole number an improper fraction, write the whole number over the number 1.

Step 2: Invert the divisor and change the ÷ symbol to a × symbol:

$$\begin{array}{ccc} \frac{6}{1} & \div & \frac{3}{4} \\ \downarrow & \downarrow & \downarrow \\ \frac{6}{1} & \times & \frac{4}{3} \end{array}$$

Step 3: Cancel and multiply across:

$$\frac{\overset{2}{\cancel{6}}}{1} \times \frac{4}{\underset{1}{\cancel{3}}} = \frac{8}{1} = 8$$

EXAMPLE 2: Divide: $24 \div \frac{7}{8}$

Solution:

Step 1: Rewrite the whole number as an improper fraction:

$$24 = \frac{24}{1}$$

Step 2: Invert the divisor and change the ÷ symbol to a × symbol:

$$\begin{array}{ccc} \frac{24}{1} & \div & \frac{7}{8} \\ \downarrow & \downarrow & \downarrow \\ \frac{24}{1} & \times & \frac{8}{7} \end{array}$$

Step 3: Multiply across:

$\frac{24}{1} \times \frac{8}{7} = \frac{192}{7}$ We cannot cancel, so simply multiply across.

Step 4: Change the improper fraction, $\frac{192}{7}$, to a mixed number:

$$\frac{192}{7} = \begin{array}{r} 27 \\ 7\overline{)192} \\ \underline{14} \\ 52 \\ \underline{49} \\ 3 \end{array} = 27\tfrac{3}{7}$$

Remember.

1. To divide fractions, invert the divisor and multiply.
2. To invert means to turn upside down, or reverse the numerator and the denominator:

$$\frac{2}{3} \text{ inverted } = \frac{3}{2}$$

$$\frac{7}{8} \text{ inverted } = \frac{8}{7}$$

$$\frac{29}{30} \text{ inverted } = \frac{30}{29}$$

Name ____________________ Date ____________________

UNIT 28 Exercises

Divide these problems. Use cancellation where possible.

1. $\frac{3}{4} \div \frac{1}{2}$
2. $\frac{5}{6} \div \frac{2}{3}$
3. $\frac{4}{5} \div \frac{1}{7}$
4. $\frac{1}{8} \div \frac{1}{9}$
5. $\frac{3}{7} \div \frac{9}{10}$
6. $\frac{7}{8} \div \frac{3}{4}$
7. $\frac{1}{3} \div \frac{5}{8}$
8. $\frac{7}{8} \div \frac{5}{8}$
9. $\frac{2}{3} \div \frac{1}{7}$

ANSWERS

1. ____________
2. ____________
3. ____________
4. ____________
5. ____________
6. ____________
7. ____________
8. ____________
9. ____________

10. $\frac{1}{5} \div \frac{1}{9}$

11. $\frac{6}{7} \div \frac{2}{5}$

12. $\frac{1}{2} \div \frac{3}{8}$

13. $\frac{7}{8} \div \frac{3}{11}$

14. $\frac{9}{10} \div \frac{2}{15}$

15. $\frac{2}{3} \div \frac{15}{16}$

16. $\frac{21}{23} \div \frac{7}{23}$

17. $\frac{11}{12} \div \frac{3}{4}$

18. $\frac{3}{8} \div \frac{5}{12}$

19. $\frac{15}{19} \div \frac{5}{38}$

20. $\frac{17}{21} \div \frac{17}{21}$

21. $\frac{1}{2} \div \frac{3}{16}$

ANSWERS

10. ________
11. ________
12. ________
13. ________
14. ________
15. ________
16. ________
17. ________
18. ________
19. ________
20. ________
21. ________

Name ______________________ Date ______________________

UNIT 28 Applications

1. How many $\frac{1}{6}$'s are there in $\frac{3}{4}$?

ANSWERS

1. ____________
2. ____________
3. ____________

2. An office worker can type an envelope in 2 minutes. How many envelopes can the worker type in $\frac{1}{2}$ hour? (Hint: 2 minutes = $\frac{2}{60}$ of an hour.)

3. David bought $\frac{3}{4}$ pound of ground beef. How many patties can he make if each patty is to weigh 4 ounces? (Hint: 4 oz. = $\frac{4}{16}$ lb.)

4. How many 2-ounce cookies can be made from $\frac{3}{4}$ pound of flour? (A pound equals 16 ounces.)

ANSWERS

4. ____________
5. ____________
6. ____________
7. ____________
8. ____________
9. ____________

5. How many 15-minute periods are there in two hours?

6. How many quarters ($\$\frac{1}{4}$) are there in $5?

7. How many $\frac{1}{2}$ gallon containers are there in 10 gallons of milk?

8. How many 10-minute periods are there in $\frac{1}{2}$ hour?

9. If 3 items cost $\$\frac{3}{4}$, what does one item cost?

UNIT 29 Dividing Mixed Numbers

DIVIDING ONE MIXED NUMBER BY ANOTHER MIXED NUMBER

To divide a mixed number by another mixed number, follow these steps:

Step 1: Change the mixed numbers to improper fractions.

Step 2: Change the ÷ symbol to a × symbol and invert the divisor.

Step 3: Cancel, where possible, and multiply across.

Step 4: Change any answers that are improper fractions to whole numbers or to mixed numbers.

EXAMPLE 1: Divide: $6\frac{1}{8} \div 4\frac{2}{3}$

Solution:

Step 1: Change the mixed numbers, $6\frac{1}{8}$ and $4\frac{2}{3}$, to improper fractions:

$$6\tfrac{1}{8} = \frac{(8 \times 6 = 48) + 1}{8} = \frac{49}{8}$$

$$4\tfrac{2}{3} = \frac{(3 \times 4 = 12) + 2}{3} = \frac{14}{3}$$

Step 2: Change the ÷ symbol to a × symbol and invert the divisor, $\frac{14}{3}$:

Remember: Bring down the first fraction exactly as it is written

$$\frac{49}{8} \div \frac{14}{3} \;\rightarrow\; \frac{49}{8} \times \frac{3}{14}$$

Step 3: Cancel and multiply across:

$$\frac{\overset{7}{\cancel{49}}}{8} \times \frac{3}{\underset{2}{\cancel{14}}} = \frac{21}{16}$$

Step 4: Change the improper fraction, $\frac{21}{16}$, to a mixed number:

$$\frac{21}{16} = 16\overline{)21} \;\; \text{quotient } 1,\; 21 - 16 = 5 \;\; = 1\tfrac{5}{16}$$

EXAMPLE 2: Divide: $3\frac{1}{2} \div 5\frac{1}{4}$

Solution: Change the mixed numbers to improper fractions, invert the divisor, cancel and multiply:

$$3\tfrac{1}{2} \div 5\tfrac{1}{4} = \frac{7}{2} \div \frac{21}{4} = \frac{\overset{1}{\cancel{7}}}{\underset{1}{\cancel{2}}} \times \frac{\overset{2}{\cancel{4}}}{\underset{3}{\cancel{21}}} = \frac{2}{3}$$

DIVIDING WHOLE NUMBERS BY MIXED NUMBERS

To divide a whole number by a mixed number (or a mixed number by a whole number), you change the whole number and the mixed number to improper fractions, invert the divisor, and multiply. Before you multiply, remember to perform any cancellation possible.

EXAMPLE 1: Divide: $24 \div 3\frac{1}{3}$

Solution:

Step 1: Change the whole number and the mixed number to improper fractions:

$$24 = \frac{24}{1}$$

$$3\frac{1}{3} = \frac{(3 \times 3 = 9) + 1}{3} = \frac{10}{3}$$

Step 2: Change the ÷ symbol to a × symbol and invert the divisor:

$$\frac{24}{1} \div \frac{10}{3}$$
$$\downarrow \quad \downarrow \quad \downarrow$$
$$\frac{24}{1} \times \frac{3}{10}$$

Step 3: Cancel and multiply across:

$$\frac{\overset{12}{\cancel{24}}}{1} \times \frac{3}{\underset{5}{\cancel{10}}} = \frac{36}{5}$$

Step 4: Change the improper fraction, $\frac{36}{5}$, to a mixed number:

$$\frac{36}{5} = \begin{array}{r} 7 \\ 5\overline{)36} \\ \underline{35} \\ 1 \end{array} = 7\frac{1}{5}$$

EXAMPLE 2: Divide: $12\frac{4}{5} \div 8$

Solution: Convert both the mixed number and the whole number to improper fractions:

$$12\frac{4}{5} = \frac{64}{5}$$

$$8 = \frac{8}{1}$$

Invert the divisor, cancel, and multiply:

$$\frac{\overset{8}{\cancel{64}}}{5} \times \frac{1}{\underset{1}{\cancel{8}}} = \frac{8}{5} = \begin{array}{r} 1 \\ 5\overline{)8} \\ \underline{5} \\ 3 \end{array} = 1\frac{3}{5}$$

Name ______________________ Date ______________________

UNIT 29 Exercises

Divide each of these problems. Simplify where possible by using cancellation.

1. $32\frac{1}{4} \div 5\frac{1}{2}$
2. $16\frac{3}{4} \div 10\frac{1}{8}$
3. $5\frac{1}{8} \div 2\frac{1}{9}$
4. $8\frac{2}{3} \div 6\frac{1}{4}$
5. $10\frac{2}{3} \div 4\frac{1}{6}$
6. $6\frac{2}{3} \div 2\frac{2}{5}$
7. $3\frac{3}{4} \div 2\frac{1}{4}$
8. $48\frac{1}{2} \div 12\frac{1}{8}$
9. $9\frac{3}{5} \div 3\frac{3}{7}$

ANSWERS

1. ______________
2. ______________
3. ______________
4. ______________
5. ______________
6. ______________
7. ______________
8. ______________
9. ______________

10. $12 \div 3\frac{1}{3}$

11. $24 \div 10\frac{2}{3}$

12. $20 \div 4\frac{2}{5}$

ANSWERS

10. ____________
11. ____________
12. ____________
13. ____________
14. ____________
15. ____________
16. ____________
17. ____________
18. ____________
19. ____________
20. ____________
21. ____________

13. $6 \div 3\frac{7}{8}$

14. $8 \div 3\frac{1}{5}$

15. $10 \div 3\frac{1}{3}$

16. $12\frac{3}{5} \div 9$

17. $12\frac{1}{2} \div 5$

18. $16\frac{1}{4} \div 6$

19. $12\frac{1}{4} \div 7$

20. $24\frac{3}{8} \div \frac{4}{5}$

21. $36\frac{2}{3} \div \frac{7}{9}$

Name ______________________ Date ______________________

UNIT 29 Applications

1. A carton of fruit juice holds $48\frac{1}{3}$ ounces. How many $6\frac{1}{4}$-ounce glasses can be filled from the carton?

ANSWERS

1. ____________
2. ____________
3. ____________

2. If a $10\frac{1}{2}$ ounce box of cereal costs \$1.00, how much does one ounce cost? (Hint: Change \$1.00 to an improper fraction: $\$\frac{1}{1}$.)

3. How many $3\frac{1}{3}$-foot sections of material can be cut from a larger section that is 50 feet long?

4. If it takes $8\frac{1}{2}$ ounces of flour to make a cake, how many cakes can be made from a 68-ounce bag of flour?

ANSWERS

4. ____________
5. ____________
6. ____________
7. ____________

5. If Ben drove 255 miles in $4\frac{1}{4}$ hours, how many miles per hour did he average?

6. How many 3-foot pieces of ribbon can be cut from a longer piece measuring $13\frac{1}{2}$ feet?

7. Larry is going to divide \$18.50 among his three children. How much is each child's share? (Hint: \$18.50 = $\$18\frac{1}{2}$.)

Name ______________________ Date ______________________

8. If a $10\frac{1}{4}$-ounce bag of peanuts is divided four ways, what would be the size of each share?

ANSWERS

8. ____________

9. ____________

10. ____________

9. Kathy's new car averages $24\frac{1}{2}$ miles to a gallon of gas. On a recent trip, she drove 392 miles. How much gas did she use?

10. After cooking, a large steak weighed $18\frac{1}{4}$ ounces. What is the size of one portion if the steak is to serve four people?

Work Space

UNIT 30 The Meaning of Decimals and Reading Decimals

DECIMALS ARE FRACTIONS

Like a common fraction, a **decimal** stands for a part of a whole unit; this is why decimals are often called **decimal fractions.** But unlike a common fraction, which can have any number as its denominator, a decimal fraction always has a denominator that is a power of 10. Powers of 10, as you recall, are 10, 100, 1,000, 10,000, 100,000, and so on. Also, the denominator of a decimal fraction is not written; it is understood to be 10, 100, 1,000, and so on.

The sign of the decimal is the decimal point. Digits written to the left of the decimal point stand for whole numbers; digits written to the right of the decimal point stand for decimal fractions, or partial units. For example, look at this number:

decimal point

2 5 5 . 1 2 5

← whole numbers | decimal fractions →

As you remember from Unit 1, the value of a digit within a whole number depends on its position, or its "place"; this is called "place value." Place value also applies to decimals, or to numbers that are not whole numbers. The following chart shows the place value of whole numbers (from ones to billions), and the place value of decimals (from tenths to billionths).

Billions	Hundred millions	Ten millions	Millions	Hundred thousands	Ten thousands	Thousands	Hundreds	Tens	Ones	**Decimal point**	Tenths	Hundredths	Thousandths	Ten thousandths	Hundred thousandths	Millionths	Ten millionths	Hundred millionths	Billionths
										.									

As you recall, each place to the left of the decimal point is worth ten times the previous place. But each place to the right of the decimal point is worth one-tenth of the place before it. Notice that places to the right of the decimal point end in "ths."

You read a decimal the same way you read a whole number, except that you only read the place value of the right-hand digit.

EXAMPLE 1: Read .87

Solution:

Read the decimal as you would a whole number: "eighty-seven"

Then, read the place value of the right-hand digit. The right hand digit, 7, is in the *hundredths* place; thus, .87 is read as "eighty-seven hundredths."

Billions	Hundred millions	Ten millions	Millions	Hundred thousands	Ten thousands	Thousands	Hundreds	Tens	Ones	**Decimal point**	Tenths	Hundredths	Thousandths	Ten thousandths	Hundred thousandths	Millionths	Ten millionths	Hundred millionths	Billionths
										.	8	7							

.87 = eighty-seven hundredths

EXAMPLE 2: Read .675
Solution:
Read the decimal as a whole number: "six hundred seventy-five." Then, read the place value of the right-hand digit. The right-hand digit, 5, is in the *thousandths* place; thus, .675 is read as "six hundred seventy-five thousandths."

Billions	Hundred millions	Ten millions	Millions	Hundred thousands	Ten thousands	Thousands	Hundreds	Tens	Ones	**Decimal point**	Tenths	Hundredths	Thousandths	Ten thousandths	Hundred thousandths	Millionths	Ten millionths	Hundred millionths	Billionths
										.	6	7	5						

.675 = six hundred seventy-five thousandths

EXAMPLE 3: Read .0002
Solution:
The right-hand digit, 2, is in the ten-thousandths place. Thus, the decimal is read as "two ten-thousandths."

Billions	Hundred millions	Ten millions	Millions	Hundred thousands	Ten thousands	Thousands	Hundreds	Tens	Ones	**Decimal point**	Tenths	Hundredths	Thousandths	Ten thousandths	Hundred thousandths	Millionths	Ten millionths	Hundred millionths	Billionths
										.	0	0	0	2					

.0002 = two ten-thousandths

Let's look at how some other decimals should be read and how they would appear as fractions.

Decimal	Read As	Fraction
.7	7 tenths	$\frac{7}{10}$
.16	16 hundredths	$\frac{16}{100}$
.003	3 thousandths	$\frac{3}{1,000}$
.0025	25 ten-thousandths	$\frac{25}{10,000}$
.000175	175 millionths	$\frac{175}{1,000,000}$

MIXED DECIMALS

A **mixed decimal** (mixed number) is a whole number and a decimal fraction—28.4, for example. To read a mixed decimal, first read the whole number, say the word "and" when you reach the decimal point, and then read the decimal.

EXAMPLE 1: Read 345.28
Solution:

Read the whole number: "three hundred forty-five;" then, read the decimal: "and twenty-eight hundredths."

345.28 = three hundred forty-five and twenty-eight hundredths

Notice that in reading this mixed decimal, we did not say the word "and" until we reached the decimal point. Remember from Unit 1 that the word "and" is not used when you read whole numbers. It is used to separate whole numbers (on the left side of the decimal point) from decimal fractions (on the right side of the decimal point).

EXAMPLE 2: Read 2,455.5
Solution:

two thousand, four hundred fifty-five and five tenths

again, notice that "and" was only used to separate the whole number from the decimal

Let's look at a few more examples.

4.003 = four and three thousandths

25.02 = twenty-five and two hundredths

44.0004 = forty-four and four ten-thousandths

175.9 = one hundred seventy-five and nine tenths

16.092 = sixteen and ninety-two thousandths

Remember.

1. A decimal is a special kind of fraction with a denominator that is a power of 10 (10, 100, 1,000, etc.).

2. The denominator of a decimal is read, but not written; it is understood:

   .5 = five-tenths = $\frac{5}{10}$

   .25 = twenty-five hundredths = $\frac{25}{100}$

   .125 = one hundred twenty-five thousandths = $\frac{125}{1000}$

3. When reading mixed decimals, the word "and" is used only to separate whole numbers from decimals.

Work Space

Name ______________________ Date ______________________

UNIT 30 Exercises

Write these decimals and mixed decimals in words.

1. .8 ______________________

2. .08 ______________________

3. .007 ______________________

4. .15 ______________________

5. .5125 ______________________

6. .00008 ______________________

7. .75431 ______________________

8. .025 ______________________

9. .0012 ______________________

10. .4444 ______________________

11. .04175 ______________________

12. .00685 ______________________

13. .999999 ______________________

14. 145.8 ______________________

15 2.00876 ______________________

16. 7.85 ______________________

17. 1.0025 ________________________________

18. 45.0002 ________________________________

19. 767.002 ________________________________

20. 2,345.02 ________________________________

21. 3,458.0521 ________________________________

22. 75,678.1 ________________________________

23. $35.42 ________________________________

24. $5,587.96 ________________________________

25. $34,582.97 ________________________________

Write each of these as a decimal fraction.

26. twenty-five hundredths ________________________

27. three tenths ________________________________

28. forty-five hundredths ________________________

29. two hundred fifty-five thousandths ________________

30. five hundred thirty-five ten-thousandths ____________

31. two and fifty-eight thousandths ________________

32. forty-four and 9 tenths ________________________

33. one and forty-four hundredths ________________

34. five tenths ________________________________

Name ______________________ Date ______________________

UNIT 30 Exercises

Write these decimals and mixed decimals in words.

1. .8 ______________________

2. .08 ______________________

3. .007 ______________________

4. .15 ______________________

5. .5125 ______________________

6. .00008 ______________________

7. .75431 ______________________

8. .025 ______________________

9. .0012 ______________________

10. .4444 ______________________

11. .04175 ______________________

12. .00685 ______________________

13. .999999 ______________________

14. 145.8 ______________________

15 2.00876 ______________________

16. 7.85 ______________________

17. 1.0025 ______________________________

18. 45.0002 ______________________________

19. 767.002 ______________________________

20. 2,345.02 ______________________________

21. 3,458.0521 ______________________________

22. 75,678.1 ______________________________

23. $35.42 ______________________________

24. $5,587.96 ______________________________

25. $34,582.97 ______________________________

Write each of these as a decimal fraction.

26. twenty-five hundredths ______________________

27. three tenths ______________________

28. forty-five hundredths ______________________

29. two hundred fifty-five thousandths ______________

30. five hundred thirty-five ten-thousandths ____________

31. two and fifty-eight thousandths ________________

32. forty-four and 9 tenths ______________________

33. one and forty-four hundredths ________________

34. five tenths ______________________

Name ______________________ Date ______________________

35. four ten-thousandths ______________________

36. seven hundred three and four thousandths ______________________

37. seventy-five and two hundred-thousandths ______________________

38. four millionths ______________________

39. ten and eighty-eight ten-thousandths ______________________

40. seventy-five millionths ______________________

Write each of these decimals as common fractions.

	Decimal	Common Fraction
41.	.05	______
42.	.8	______
43.	.45	______
44.	.125	______
45.	.905	______
46.	.0125	______
47.	.25	______
48.	.025	______
49.	.775	______
50.	.0004	______

Write these common fractions as decimals.

51. $\frac{3}{10}$ 52. $\frac{4}{100}$ 53. $\frac{9}{10}$ 54. $\frac{7}{10}$ 55. $\frac{6}{100}$ 56. $\frac{9}{100}$ 57. $\frac{45}{100}$

58. $\frac{75}{1,000}$ 59. $\frac{225}{10,000}$ 60. $\frac{19}{100,000}$

ANSWERS

51. ________
52. ________
53. ________
54. ________
55. ________
56. ________
57. ________
58. ________
59. ________
60. ________

Name ______________________ Date ______________________

UNIT 30 Applications

ANSWERS

1. ______
2. ______
3. ______
4. ______
5. ______
6. ______
7. ______

1. Out of each $100 of operating expenses, Rodriguez Products Company pays $60 in salaries and wages. Write as a decimal fraction the part of operating expenses spent on salaries and wages.

2. Stern Company reported a one-tenth increase in profits over a year ago. Write the increase as a decimal fraction.

3. During March, McGram Company placed 1,000 units of a certain product into production. But because of spoilage, only 975 units were usable. Write the number of units usable as a decimal fraction of the original amount.

4. Champe Summers paid $1,000 down on the purchase of a $10,000 car. Write the down payment as a decimal fraction of the price of the car.

5. Jane McKinney, a sales representative for a candy company, receives a $20 commission on each $100 she sells. Write this as a decimal fraction.

6. Tim McEwen got a raise from $8 an hour to $10 an hour. Write the first wage as a decimal fraction of the new one.

7. A sports coat marked for sale for $100 was reduced by $20. Write the reduction as a decimal fraction.

8. In a recent survey, 7 out of 10 people said that they prefer to live in a warm climate. Write this as a decimal fraction.

9. Write these as decimal fractions:
   a. 6 out of 10

   b. 8 out of 100

   c. 12 out of 1,000

   d. 300 out of 1,000

   e. 20 out of 10,000

   f. 5,000 out of 10,000

ANSWERS

8. ____________

9. a. ____________
   b. ____________
   c. ____________
   d. ____________
   e. ____________
   f. ____________

UNIT

31

Rounding Off Decimals

In Unit 2, we rounded off whole numbers to a given place, such as tens, hundreds, thousands, and so on. We learned that rounding off numbers is a way to make the numbers easier to work with and easier to remember.

Decimals, like whole numbers, can be rounded off to make them easier to work with and easier to remember. When rounding off decimals, you follow the same steps you used to round off whole numbers, except that digits to the right of the place being rounded are dropped (rather than being changed to zeros). Let's see how this is done.

To round a decimal, you:

Step 1: Designate the place to which the decimal is to be rounded (we will call this digit the **rounding place**).

Step 2: Underline the rounding place.

Step 3: Look only at the digit to the immediate right of the rounding place; if this digit is 5 or more, increase the digit in the rounding place by 1; if this digit is 4 or less, do not change the digit in the rounding place.

Step 4: Drop all digits to the right of the rounding place.

EXAMPLE 1: Round .875 to the nearest tenth.

Solution:

Step 1: The place to which the decimal is to be rounded is tenths (remember, the tenths place is one place to the right of the decimal point).

↓ rounding place

.875

Step 2: Underline the rounding place:

$.\underline{8}75$

Step 3: Look at the digit to the immediate right of the rounding place; it is 7, which is 5 or more. So, the digit in the rounding place is increased by 1:

$\overset{9}{\not{8}}75 = .9$

The digit to the right of the rounding place is 7; increase the rounding place by 1: 8 + 1 = 9.

Step 4: Drop all digits to the right of the rounding place:

$.\overset{9}{\not{8}}\not{7}\not{5} = .9$

EXAMPLE 2: Round .6745 to the nearest hundredth.

Solution:

.6745

↑ The rounding place is hundredths.

.6<u>7</u>45

↑ The digit to the right of the rounding place is less than 5; do not change the number in the rounding place.

.67~~45~~ = .67

↑ Drop all digits to the right of the rounding place.

Let's look at a few more examples.

25.7881 rounded to the nearest hundredth = 25.79

$325.125 rounded to the nearest cent (hundredth) = $325.13

75.999 rounded to the nearest tenth = 76.0

Remember. When rounding decimals, look *only* at the digit to the immediate right of the rounding place; **do not consider any other digit.**

Name ______________________ Date ______________________

UNIT 31 Exercises

Round these numbers as shown. The first one is done as an example.

Decimal	Tenths	Hundredths	Thousandths
1. .5984	.6	.60	.598
2. .7775			
3. .8011			
4. .8275			
5. .97347			
6. .65216			
7. 1.25			
8. 4.775			
9. .0685			
10. .222222			
11. .66666666			
12. 17.8962			
13. 275.9951			
14. 2,455.9875			
15. 15,977.9999			

Round each of these to the nearest tenth (1 place).

16. .075
17. .8875
18. .09675
19. .6682
20. .9999
21. 45.666
22. 745.0125
23. 238.9025
24. 2,575.88217
25. 12,455.89165

Round these to the nearest hundredth (2 places).

26. .027
27. .9726
28. .9012
29. .2349
30. .4567
31. 45.6668
32. 76.4582
33. 245.9999
34. 1,259.08212
35. 12,000.345

Round these to the nearest thousandths (3 places).

36. .03877
37. .329767
38. .08254
39. .49875
40. .098714
41. .003872
42. .93276
43. .97254
44. .1287945
45. 12.3457

ANSWERS

16. ____________
17. ____________
18. ____________
19. ____________
20. ____________
21. ____________
22. ____________
23. ____________
24. ____________
25. ____________
26. ____________
27. ____________
28. ____________
29. ____________
30. ____________
31. ____________
32. ____________
33. ____________
34. ____________
35. ____________
36. ____________
37. ____________
38. ____________
39. ____________
40. ____________
41. ____________
42. ____________
43. ____________
44. ____________
45. ____________

Name ______________________ Date ______________________

Round these to the nearest ten-thousandths (4 places).

46. .97563
47. .99999999
48. .29836
49. .0387654
50. .987654
51. .12978678
52. 2.498766
53. 245.69265
54. 98.00345
55. 2,345.98252

Round these to the nearest hundred-thousandths (5 places).

56. .1879768
57. .218096
58. .15678656
59. .1879661
60. .18767867
61. .1781567
62. .21798675
63. .28977891
64. 567.378946
65. 451.789419

Round these to the nearest millionths (6 places).

66. .0078654
67. .7886543
68. .28072687
69. .45870936
70. .21467389
71. 1,238.1875462
72. 5,296.5487611
73. 71.0000876
74. 6,478.4376193
75. 12,345.6715438

ANSWERS

46. ____________
47. ____________
48. ____________
49. ____________
50. ____________
51. ____________
52. ____________
53. ____________
54. ____________
55. ____________
56. ____________
57. ____________
58. ____________
59. ____________
60. ____________
61. ____________
62. ____________
63. ____________
64. ____________
65. ____________
66. ____________
67. ____________
68. ____________
69. ____________
70. ____________
71. ____________
72. ____________
73. ____________
74. ____________
75. ____________

Work Space

Name ______________________ Date ______________________

UNIT 31 Applications

1. When figuring the total amount to charge a customer, a wholesaler arrived at a figure of $245.6865. Round this figure to even cents (hundredths).

2. A scoop of jelly beans weighs 2.78 pounds. What is the weight of the candy rounded to the nearest tenth?

3. Round 12.4598376 to the nearest (a) ten-thousandth, (b) thousandth, (c) hundredth, (d) tenth.

4. The thickness of a metal rod is .8245 inch. Round the measurement to the nearest hundredth.

5. The value of an ounce of gold is $382.525. Round this figure to even cents.

6. The exact distance from Tom's house to his job is 2.458 miles. Round this figure to the nearest (a) tenth of a mile and (b) hundredth of a mile.

ANSWERS

1. ________
2. ________
3. a. ________
   b. ________
   c. ________
   d. ________
4. ________
5. ________
6. a. ________
   b. ________

7. The label on a package of ground beef shows that the total weight of the package is 2.75 pounds. Round the weight off to the nearest tenth of a pound.

ANSWERS

7. ____________

8. Following are wholesale prices on items sold by Englewood Company. Round each price to even cents.

Item	Price	
B40	$4.567	________
D18	$1.019	______
A10	$0.958	________
H14	$6.666	________
A19	$3.458	________
H10	$0.999	________

Work Space

UNIT 32

Comparing Decimals

We have learned that the terms of a common fraction can be changed without changing the value of the fraction. For example, we can raise the terms of a common fraction by multiplying both terms by the same nonzero whole number. We can also reduce the terms of a common fraction by dividing both terms by the same number.

The terms of common fractions are changed to make them easier to work with and easier to compare. You can also make decimals easier to work with and easier to compare by changing them so that they will have a common denominator. Decimals have a common denominator when they have the *same number of decimal places*. So, .02 (two hundredths) and .45 (forty-five hundredths) have a common denominator of 100. In the same way, the decimals .005 and .125 have a common denominator of 1,000. Let's look at some examples that compare decimals.

EXAMPLE 1: Which decimal has a larger value, .385 or .39?

Solution:

Step 1: Rewrite the decimals with the same number of places:

$$.385 = .385$$

$$.39 = .390$$

Step 2: Compare the results (the higher number will have the larger value):

The higher decimal is .390, so it has a larger value.

EXAMPLE 2: Which decimal is smaller, .8 or .08?

Solution:

Step 1: Rewrite the decimals with the same number of places:

$$.8 = .80$$

$$.08 = .08$$

Step 2: Look at the results:

$$.80 = \frac{80}{100}$$

$$.08 = \frac{8}{100}$$

.08 is smaller because it is 8 parts out of 100, while .80 is 80 parts out of 100.

.08 is smaller than .8.

From these examples we can see that writing zeros to the right of the decimal point does not change the value of the decimal; it simply changes the denominator. As another example, let's take .5. If a zero is added, it becomes .50; if another zero is added, it becomes .500, and so on. These are all ways of writing the same value. We can see this more clearly if we rewrite the decimal as a common fraction and reduce it to lowest terms:

$$.5 = \frac{5}{10} = \frac{1}{2}$$

$$.50 = \frac{50}{100} = \frac{1}{2}$$

$$.500 = \frac{500}{1{,}000} = \frac{1}{2}$$

.5, .50, .500 all reduce to $\frac{1}{2}$, thus they have the same value.

Let's look at another example.

EXAMPLE: Show that .2 and .20 have the same value.
Solution:

$$.2 = \frac{2}{10} = \frac{1}{5}$$

$$.20 = \frac{20}{100} = \frac{1}{5}$$

We can also use models to show that writing zeros after the last digit in a decimal doesn't change its value. The following models show that .3 has the same value as .30:

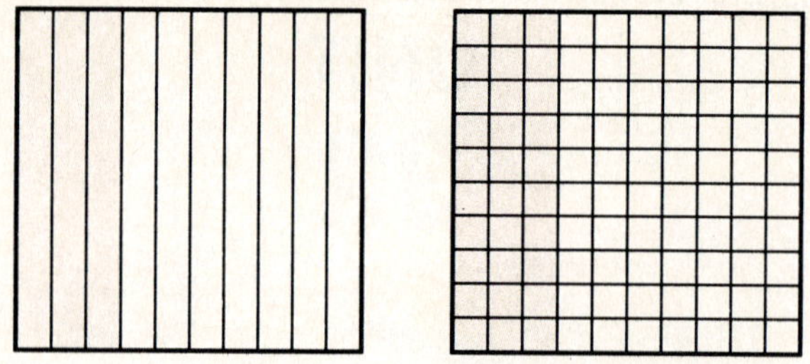

3 tenths	=	30 hundredths
.3	=	.30

To add clarity, zeros can also be written to the left of the decimal point. For example, .58 can be written as 0.58. This is a way to show that there are no whole numbers with the decimal. You often see this when dollars and cents are written. To express \$.69, for example, you can write \$0.69. Writing it this way states definitely that there are no whole dollars, just the 69 cents.

Name ______________________ Date ______________________

UNIT 32 Exercises

In each of these problems, circle the decimal with the largest value.

1. .05 .005 .0005
2. .04 .040 .004
3. .70 .07 0.07
4. .8 .808 0.008
5. .55 .055 0.50
6. .4 .40 .400
7. .9 .995 .095
8. .02 .20 .020

Prove that the following decimals are of equal value.

9. .6 .60
10. .8 .80 .800
11. .90 .900 0.9
12. .40 .4000

13. .009 0.009

14. .70 0.7 .700

In problems 15–20, underline the decimal with the largest value.

15. .8 .08 .085

16. .65 .665 .065

17. .225 .3 .68

18. .002 .200 0.002

19. .35 .305 0.305

20. .6 .06 .60

In problems 21–24, underline the decimal with the smallest value.

21. .02 .002 .0002

22. .004 .04 0.0004

23. .9 .99 0.99

24. .995 .990 .900

Name ______________________ Date ______________________

UNIT 32 Applications

1. On Tuesday, Reggi jogged .8 miles and Colin jogged .75 miles. Who jogged farther? How much farther?

2. Which is larger, eight-tenths of a dollar or three-fourths of a dollar?

3. Part of Dominick's science project was to compare the weights of two test tubes of a heavy liquid metal. One weighed .84 pound and the other weighed .9 pound. Which one weighed more?

4. Harold's height is 5.8 feet, while Sam's height is 5.64 feet. Who is taller?

5. Which is larger, .095 or .0925?

6. A bag of Brand A sunflower seeds holds 12.2 ounces, while a bag of Brand B sunflower seed holds 12.25 ounces. If the cost of each is the same, which is the better value?

7. Which is smaller, .020 or .022?

8. Prove that .8 and .800 have the same value.

ANSWERS

1. ____________
2. ____________
3. ____________
4. ____________
5. ____________
6. ____________
7. ____________

Work Space

UNIT
33 Adding and Subtracting Decimals and Mixed Numbers

ADDING DECIMALS AND MIXED NUMBERS

Adding decimals is very similar to adding whole numbers. When adding decimals, however, you must line up the decimal points in a straight line. By lining up the decimal points, you automatically line up the place values of the problem. Use this rule to help you when adding decimals and mixed decimals:

> **RULE**
> To add decimals, line up the decimal points in a vertical line, then add, following the rules for addition of whole numbers.

EXAMPLE 1: Add .521 + .455
Solution:

line up the decimal points

```
   .521
 + .455
   .976
```

place decimal point here

EXAMPLE 2: Add 225.58 + 14.25

Solution:

line up the decimal points

```
 225.58
  14.25
 239.83
```

place decimal point here

If you are adding decimals that do not have the same number of places, you can make the problem easier to work with by filling in empty spaces with zeros (adding zeros to the right of the decimal point does not change the value of the number).

EXAMPLE 1: Add .02 + .564 + .233
Solution:

```
 .02      .020  ← fill in empty space with a zero
 .564  =  .564
 .233     .233
          .817
```

EXAMPLE 2: Add 2.99 + .7 + .821

Solution:

$$\begin{array}{r} 2.99 \\ .7 \\ \underline{.821} \end{array} \quad = \quad \begin{array}{r} 2.990 \\ .700 \\ \underline{.821} \\ 4.511 \end{array}$$

fill in empty spaces with zeros

SUBTRACTING DECIMALS AND MIXED DECIMALS

To subtract decimals and mixed decimals, you set up the problem as you did for addition; that is, line up the decimal points in a vertical line, and subtract as you would with whole numbers.

EXAMPLE 1: Subtract: 48.75 − 24.61
Solution:

line up the decimal points

$$\begin{array}{r} 48.75 \\ -\ \underline{22.61} \\ 26.14 \end{array}$$

place decimal point here

Check: Remember, to check a subtraction problem, add the difference (the answer) to the subtrahend (the second number); the sum will equal the minuend (the top number).

$$\begin{array}{r} 26.14 \\ +\ \underline{22.61} \\ 48.75 \end{array}$$

EXAMPLE 2: Subtract: 345.52 − 31.5
Solution:
Line up the decimal points correctly:

$$\begin{array}{r} 345.52 \\ -\ \underline{31.5} \end{array}$$

Fill in empty space with a zero and bring the decimal point into the answer:

$$\begin{array}{r} 345.52 \\ -\ \underline{31.50} \end{array}$$

fill in empty space

Subtract as with whole numbers:

$$\begin{array}{r} 345.52 \\ -\ \underline{31.50} \\ 314.02 \end{array}$$

Check:

$$\begin{array}{r} 314.02 \\ +\ \underline{31.50} \\ 345.52 \end{array}$$

EXAMPLE 3: Subtract: 78.51 − 12.39
Solution:

$$\begin{array}{r} {\scriptstyle 4\ 1} \\ 78.\not{5}1 \\ -\ \underline{12.39} \\ 66.12 \end{array}$$

remember to borrow, because you cannot subtract 9 from 1

Check:

$$\begin{array}{r} 66.12 \\ +\ \underline{12.39} \\ 78.51 \end{array}$$

We have stressed that zeros can be added to the right of the decimal point without changing the value of the decimal number. If, in a subtraction problem, the minuend (the top number) has fewer places than the subtrahend (the bottom number), you *must* add zeros to allow you to subtract. For example, how would you subtract 4.2 − .375?

First, set up the problem with the decimal points lined up.

$$\begin{array}{r} 4.2 \\ -\ \underline{.375} \end{array}$$

Then, you would fill in the empty spaces with zeros, borrow, and subtract.

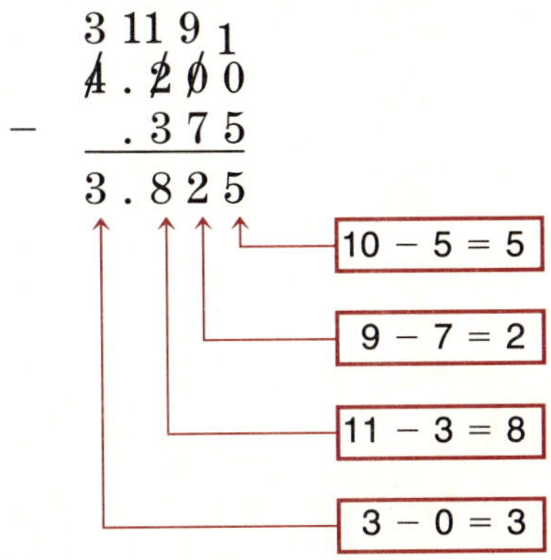

Let's look at another example.

EXAMPLE: Subtract: 5.25 − .0175
Solution:
Set up the problem correctly:

$$\begin{array}{r} 5.25 \\ -\ \underline{.0175} \end{array}$$

Fill in empty spaces with zeros and subtract:

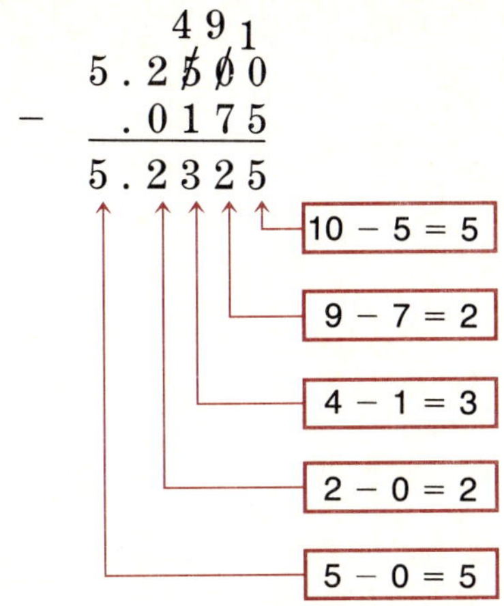

Check:

$$\begin{array}{r} 5.2325 \\ +\ \underline{.0175} \\ 5.2500 \end{array}$$

Remember. When adding and subtracting decimal numbers, the numbers are arranged in columns with the decimal points directly under one another.

Work Space

Name ____________________ Date ____________________

UNIT 33 Exercises

Add each of these sets of numbers.

1. .2 + .23 + .456
2. .24 + .09 + .34
3. .34 + .381 + .015
4. .00125 + .5812
5. .237 + .3867 + .128
6. .37 + .276 + .8756
7. .8749 + .2 + .47
8. .9215 + .04 + .0004

ANSWERS

1. ____________
2. ____________
3. ____________
4. ____________
5. ____________
6. ____________
7. ____________
8. ____________

9. .2458 + .38765 + .9

10. .125 + .78 + .025

ANSWERS

9. ____________
10. ____________
11. ____________
12. ____________
13. ____________
14. ____________
15. ____________
16. ____________

11. 2.5 + .004 + .9854

12. 12.58 + 4.589 + 2.983

13. .2 + 1.2 + 12.875

14. 1.99 + 2.3 + 4.7

15. 5.4 + .004 + .0009

16. 12.5 + 25.09 + 67.801

Name ______________________ Date ______________________

17. 125.8 + 17.51 + 2.0025

18. .00002 + 15 + 2.5

ANSWERS

17. ____________
18. ____________
19. ____________
20. ____________
21. ____________
22. ____________
23. ____________
24. ____________

19. 33.89 + 125.01 + .8 + 1.25

20. 65.8 + 21.89 + 234.7

21. $45.91 + $.59

22. $1,234.56 + $34.50

23. $.69 + $2.34 + $235.98

24. $564.89 + $1,245.89

25. \$75,617.84 + \$21,319.45 + \$12.58 + \$125.42

ANSWERS

25. ______
26. ______
27. ______
28. ______
29. ______
30. ______
31. ______
32. ______
33. ______
34. ______

Subtract each of the following problems.

26. 1.2 − .0125 27. 3.23 − .04 28. .88 − .09

29. 2.75 − 1.65 30. 12.58 − 4.75 31. 35.6 − 10.25

32. 4.675 − 3.789 33. 45.328 − 21.145 34. 8 − 2.5

Name ______________________ Date ______________________

35. .875 − .00125

36. 32 − .0092

37. 4.88 − .995

38. 125.87− 12.365

39. 458.45 − 4.675

40. 4.59 − .075

41. 15 − 12.45

42. 30 − 5.1025

43. 21.5 − 10.475

44. 55 − 12.65

45. 225 − 21.62

46. 65 − 45.75

ANSWERS

35. ____________
36. ____________
37. ____________
38. ____________
39. ____________
40. ____________
41. ____________
42. ____________
43. ____________
44. ____________
45. ____________
46. ____________

Subtract the following problems.

ANSWERS

47. ____________
48. ____________
49. ____________
50. ____________
51. ____________
52. ____________
53. ____________
54. ____________

47. 2.8 from 3.45

48. 4.785 from 8.9

49. 10.5 from 12.25

50. 18.03 from 42.005

51. 23.0075 from 55.8

52. 8.995 from 12

53. 45.87 from 778.276

54. 2.5 from 15

Name ____________________ Date ____________________

55. 17.00005 from 50

56. 2.675 from 6.5

ANSWERS

55. __________
56. __________
57. __________
58. __________
59. __________
60. __________

57. \$25.75 from \$100

58. \$.69 from \$5

59. \$21.58 from \$65

60. \$2.38 from \$10

Work Space

Name ______________________ Date ______________________

UNIT 33 Applications

1. The Simmons Company paid these bills for the week ending March 12:

Rent	$600.00
Utilities	921.43
Salaries	678.91
Miscellaneous	97.59

What was the total amount paid for expenses?

ANSWERS

1. ______________
2. ______________
3. ______________

2. On March 12, the bank account of the Hume Fitzsimmons Company had a balance of $4,567.87. On the same date, the company's bookkeeper made a deposit of $3,546.18 and wrote two checks, one for $456.98 and the other for $346.98. What was the balance of the account after these transactions?

3. Joyce jogged 6.5 miles on Tuesday, 6.8 miles on Wednesday, and 4.75 miles on Friday. How many miles did she jog in the week?

4. Bill's salary for the week was $345.85. But his employer deducted these amounts:

Federal income taxes	$38.10
State income taxes	12.45
Social security taxes	24.48
Health insurance	14.67

What was Bill's salary after all deductions were taken out?

ANSWERS

4. ____________

5. ____________

6. a. ____________

   b. ____________

5. Technical Products, Inc. had sales of $1,234,568.89 for the second quarter of this year. Expenses of $832,456.35 were reported for the same period. What was the amount of profit earned by the company? (Hint: Sales − expenses = net profit.)

6. The Fabric Store bought 125 yards of a certain fabric. On Saturday, the following separate sales of the fabric were made: 10.5 yards, 25.8 yards, and 5.4 yards. (a) How many yards of the fabric were sold, and (b) how many yards of the fabric remain in stock?

Name ______________________ Date ______________________

7. During a special sale, the Jones family bought a stereo for $499.95 and a TV set for $699.99. They made a down payment of $150 and agreed to pay the balance in installments. (a) What was the total amount of the purchase, and (b) what amount is owed after the down payment?

ANSWERS

7. a. __________

b. __________

8. __________

8. Gregg Blair bought the following books for this semester:

History	$39.95
Business math	12.45
Science	24.89
English	29.99

If Gregg had $125 with him to buy the books, how much did he have left after he bought them?

9. At 8 o'clock, Kathy and Bob set out on a hiking trail that was 29.5 miles long. After two hours, the couple had traveled 6.8 miles. How much farther did they have to go?

ANSWERS

9. ____________

10. ____________

10. On Tuesday, Sam Davis sold 50 shares of MIT stock for $1,200. What was his profit on the sale if he had earlier bought the stock for $879.97?

Work Space

UNIT

34 Multiplying Decimals

There are many real world business applications that call for the multiplication of decimal numbers. For example, if you want to buy 3 pounds of ground beef, what would be the total cost if one pound costs $1.99? In the same way, what would be your total earnings if you worked 40 hours last week and your hourly wage is $6.50? These situations, along with countless thousands of others each business day, call for decimal numbers to be multiplied.

To multiply numbers that have decimal points, follow these steps:

Step 1: Set up the problem and multiply the numbers as if they were whole numbers; that is, ignore the decimal points when you are multiplying.

Step 2: Count the number of decimal places to the right of the decimal point in the multiplicand **and** the multiplier.

Step 3: Starting from the right-hand digit in the product, move the decimal point to the left the number of places you counted in Step 2, and place the decimal point in this position; this is called **pointing off places.**

EXAMPLE 1: Multiply: 12.25 × 6.5

Solution:

Set up the problem and multiply as if the numbers were whole numbers:

```
   12.25
×   6.5
   6125
  7350
  79625
```

Count the number of decimal places to the right of the decimal point in the multiplicand and the multiplier:

```
12.25 ← 2 places to the right of the decimal point
  6.5 ← 1 place to the right of the decimal point
 6125   3 places
7350
79625
```

Since the multiplicand and the multiplier have 3 places to the right of the decimal point, the product must have 3 places to the right of the decimal point. Move the decimal point 3 places to the left in the product (point off 3 places):

```
  12.25
   6.5
  6125
 7350
 79.625 ← Start moving here.
   ↑ Place decimal point here.
```

EXAMPLE 2: Multiply: 24.25 × 25

Solution:

```
  24.25 ← two decimal places in multiplicand
×   25
 12125
 4850
 606.25 ← Point off two decimal places in product.
```

Answer = 606.25

When multiplying some decimal numbers, you will have to use zeros in the product to give you the appropriate number of decimal places.

EXAMPLE 1: Multiply .28 × .04
Solution:

```
   .28 ← 2 decimal places in multiplicand
×  .04 ← 2 decimal places in multiplier
 .0112 ← 4 decimal places must be in product
```

A zero must be placed in the product to give you 4 decimal places.

Answer = .0112

EXAMPLE 2: Multiply 6.08 × .0005
Solution:

```
   6.08 ← 2 decimal places in multiplicand
× .0005 ← 4 decimal places in multiplier
.003040 ← 6 decimal places must be in product
```

Two zeros must be placed in the product.

Answer = .003040

Remember. When multiplying decimal numbers, do the multiplication first; then, point off as many decimal places in the product as there are places in both the multiplicand and the multiplier combined.

Name ______________________ Date ______________________

UNIT 34 Exercises

Multiply each of thse problems.

1. 225 × .09
2. 375 × .25
3. 90 × .05
4. 850 × 2.5
5. 127 × .2
6. 18.5 × 12
7. 176 × .09
8. 436 × 5.8
9. 593 × .42
10. 2,487 × 3.5
11. 2,398 × .88
12. 4,597 × .075

ANSWERS

1. ______
2. ______
3. ______
4. ______
5. ______
6. ______
7. ______
8. ______
9. ______
10. ______
11. ______
12. ______

13. 25.38
12

14. 45.87
45

15. 125.6
3.4

16. 30.05
.09

17. 236.87
.14

18. 125.854
2.41

19. 345.9
.075

20. 25.9
.03

21. .125
.008

22. .458
.11

23. .975
.072

24. .0125
.08

25. 34.58
6.5

26. 56.4
.075

27. 2.87
.75

28. .458
.28

ANSWERS

13. ____________
14. ____________
15. ____________
16. ____________
17. ____________
18. ____________
19. ____________
20. ____________
21. ____________
22. ____________
23. ____________
24. ____________
25. ____________
26. ____________
27. ____________
28. ____________

Name ______________________ Date ______________________

29. .185
 .22

30. 265.9
 .601

31. 5.7632
 3.45

32. 245.8
 1.8

33. 345.07
 .125

34. 498.12
 3.78

35. 456.18
 .0009

36. 873.006
 2.05

37. 345.79
 .125

38. 984.91
 3.75

39. 2,736.5
 .08

40. 3,456.7
 12

ANSWERS

29. ________
30. ________
31. ________
32. ________
33. ________
34. ________
35. ________
36. ________
37. ________
38. ________
39. ________
40. ________

41. $24.75
×   15

42. $35.89
×   75

43. $87.12
×   90

44. $568.56
×   50

ANSWERS

41. ______________
42. ______________
43. ______________
44. ______________
45. ______________
46. ______________
47. ______________
48. ______________
49. ______________
50. ______________
51. ______________
52. ______________
53. ______________

45. $567.38
×   125

46. $37.125
×   980

47. $91.475
×   525

48. 3,800 × $.79

49. 4,582 × $.58

50. 500 × $1.95

51. 8,385 × $.09

52. 10.5 × $1.25

53. 8.25 × $.29

Name ______________________ Date ______________________

54. 5,118 × \$.30   55. 12.75 × \$1.49

ANSWERS

54. ________

55. ________

Work Space

Name ______________________ Date ______________________

UNIT 34 Applications

1. Sharon House bought 4.5 pounds of chopped meat at a cost of $1.29 a pound. How much did she pay for the meat?

ANSWERS

1. ____________
2. ____________
3. ____________

2. Pylant Products, Inc. sold 4,000 pounds of chemical lawn food at $.99 a pound. What was the total amount of the sale?

3. David Valdez earns $5.78 an hour. Last week he worked a total of 39 hours. What was his salary for the week?

4. What is the total cost of 5 cans of tuna fish, if one can costs $1.12?

ANSWERS

4. ____________

5. ____________

6. ____________

7. ____________

5. A local rental car agency advertises that their rates are $25 a day plus $.10 a mile. What would be the total cost of a two-day rental in which a total of 421 miles was driven?

6. The Center Company is going to recarpet their main office. The office is 556.5 square yards and the cost of the carpet is $12.95 a square yard. What will be the total cost of the carpet?

7. If one canister of baking powder weighs 246.8 grams, what would 5 canisters weigh?

Name ______________________ Date ______________________

8. Tom is considering buying a VCR that is advertised for $50 down with the balance payable in 12 monthly installments of $30 each. What would be the total cost of the VCR?

ANSWERS

8. ________

9. ________

10. ________

9. Katherine Kaye works as a salesperson for Swift Food Products Company. She receives a commission of $.05 per pound on all pork products she sells. Last week, she sold 2,500 pounds of pork. What was her commission on this product?

10. On a day when Coca Cola stock was trading for $78\frac{1}{4}$ ($78.25 per share), Sam McNiel bought 50 shares. What was the total amount of his purchase?

Work Space

UNIT
35

Multiplying Decimals Using Shortcuts

In earlier units, we learned that there are certain shortcuts that can save you time and increase your accuracy when multiplying whole numbers. There are also certain shortcuts that can save you time and increase your accuracy when multiplying decimal numbers. In this unit, we are going to look at some common shortcuts to multiplying decimal numbers.

MULTIPLYING DECIMAL NUMBERS BY A POWER OF 10

Remember that the powers of 10 (or multiples of 10) are 10, 100, 1,000, 10,000, 100,000, and so on. To multiply a number that has a decimal point by a power of 10, follow these two steps:

> Step 1: Count the number of zeros in the multiplier (the power of ten will be the multiplier).
>
> Step 2: Move the decimal point in the multiplicand the same number of places to the right as there are zeros in the multiplier.

EXAMPLE 1: Multiply: 78.25 × 100
Solution:

78.25 × 100 = 78.25 = **7,825**

Two zeros in the multiplier

Move the decimal point *two* places to the right.

EXAMPLE 2: Multiply 28.4 × 1,000
Solution:

28.4 × 1,000 = 28.400 = **28,400**

Three zeros in the multiplier

Move the decimal point *three* places to the right.

Notice in Example 2 that we had to add two zeros to the product as placeholders to allow us to move the decimal point the correct number of places to the right. Let's look at a few additional examples.

356.71 × 10 = 356.71 = **3,567.1**
(move 1 place to the right)

2.4 × 100 = 2.40 = **240**
(move 2 places to the right)

245.88 × 10,000 = 245.8800 = **2,458,800**
(move 4 places to the right)

MULTIPLYING BY 25 AND 50

To multiply a number by 25, first multiply by 100 (move the decimal point two places to the right), then divide by 4. For example, to multiply 24.48 by 25:

24.48 × 100 = 24.48

2,448 ÷ 4 = **612**

To multiply by 50, first multiply by 100, and then divide by 2. For example, to multiply 144.8 by 50:

$$144.8 \times 100 = 144.80$$

$$14{,}480 \div 2 = 7{,}240$$

Remember. To multiply a decimal number by a power of 10, move the decimal point in the multiplicand the same number of places to the right as there are zeros in the multiplier.

Name ______________________ Date ______________________

UNIT 35 Exercises

Multiply each of thse problems in your head.

1. 24.5 × 10
2. 45.87 × 100
3. .88 × 10
4. 25.05 × 10
5. 55.8 × 100
6. .8 × 1,000
7. .025 × 100
8. 1.2 × 1,000
9. .67 × 10
10. .002 × 1,000
11. 245.6 × 100
12. 24.09 × 10
13. 1,255.8 × 10
14. $.02 × 1,000
15. $1.55 × 100
16. $.03 × 10,000
17. $125.00 × 10
18. $2.05 × 1,000
19. $145.56 × 100
20. $.125 × 1,000
21. $1.25 × 1,000,000
22. $.375 × 100

Using shortcuts, multiply each of these problems by 25.

23. 448.48
24. 346.16
25. 2,516.32
26. $1,648.64
27. $1,216.36
28. $440.40

ANSWERS

1. ______
2. ______
3. ______
4. ______
5. ______
6. ______
7. ______
8. ______
9. ______
10. ______
11. ______
12. ______
13. ______
14. ______
15. ______
16. ______
17. ______
18. ______
19. ______
20. ______
21. ______
22. ______
23. ______
24. ______
25. ______
26. ______
27. ______
28. ______

Using shortcuts, multiply each of these problems by 50.

29. 252.10   30. 504.62   31. 290.64

32. $.44   33. $1.48   34. $2,464.22

ANSWERS

29. ____________
30. ____________
31. ____________
32. ____________
33. ____________
34. ____________

Using shortcuts, multiply these problems. Do them in your mind where possible.

Multiplicand	Multiplier	Product
35. 224.8	25	________
36. 464.2	50	________
37. 2,488.26	50	________
38. 6,822.28	25	________
39. 640.2	25	________
40. 677.8	10	________
41. 2.45	1,000	________
42. 6.789	100	________
43. 36.8	10,000	________
44. $345.68	100	________
45. $.08	100,000	________

Name ______________________ Date ______________________

Multiplicand	Multiplier	Product
46. $2.45	10,000	________
47. $1.375	100,000	________
48. $.125	1,000,000	________
49. $4,569.02	1,000	________
50. $331.09	10,000	________

Work Space

Name ______________________ Date ______________________

UNIT 35 Applications

ANSWERS

1. ____________
2. ____________
3. ____________
4. ____________
5. ____________

1. The Lennox Products Company bought 10,000 pounds of a certain raw material. If the cost per pound was $.18, what was the total cost of the raw material?

2. Find the price of 100 pounds of ground beef if one pound costs $1.99.

3. The River Road School bought 1,000 pencils at a cost of $.08 each. How much did all the pencils cost?

4. Robert Dodd gets a $.03 sales commission for each pound of meat he sells. What would be his commission in a month when he sells 10,000 pounds of this product?

5. Charter Bookkeeping Service bought 10 word processors at a cost of $948.50 each. What was the total amount paid for the word processors?

6. A train averaging 124.8 miles per hour will travel how far in 25 hours?

ANSWERS

6. ______________
7. ______________
8. ______________
9. ______________
10. ______________

7. If Dave Alvarez earns $8.55 an hour, how much will he make in 10 hours?

8. What is the cost of 50 cases of cola if one case costs $2.24?

9. Susan Lane is a sales representative for a major book publisher. Susan is paid $.255 for each mile she drives her personal car. Last year, she drove a total of 10,000 miles on company business. How much was she paid for mileage?

10. If a bottle of Vitamin E costs $6.95, what is the cost of 1,000 bottles?

UNIT

36

Dividing Decimals

You divide decimal numbers the same way you divide whole numbers. However, you must determine where to place the decimal point in the quotient (remember, the quotient is the answer in a division problem). Where you place the decimal point in the quotient depends on whether your divisor is a whole number, or whether it is a decimal. Let's look at both situations.

THE WHOLE NUMBER DIVISOR

When the divisor is a whole number, and the dividend has a decimal point, you place the decimal point in the quotient directly above the decimal point in the dividend. Then, divide as you would with whole numbers.

EXAMPLE: Divide: 71.68 by 8
Solution:
Move the decimal point into the quotient:

```
     .
8 ⟌73.68
```

The decimal point is written directly above the decimal point in the dividend.

Divide as usual:

```
    9.21
8 ⟌73.68
   72
    1 6
    1 6
       8
       8
       0
```

Notice in this example that the numbers divided evenly, that is, there was no remainder. When the numbers do not divide evenly (there is a remainder), there are two ways to handle the remainder:

1. State the remainder as a fraction, as we did in Unit 13 when we were dividing whole numbers; or
2. Identify a position to which the decimal is to be rounded; add enough zeros to the dividend to allow you to carry the division to one place beyond the rounding place; then, round off. This is the preferred procedure in business. Let's see how it works.

EXAMPLE: Divide: 76.8 by 18 and round the quotient correct to the nearest hundredth (remember, hundredths is 2 decimal places to the right of the decimal point).

Solution:

Set up the problem and move the decimal point into the quotient:

$$18\overline{)76.8}$$

Add enough zeros to the dividend to allow you to carry the division to one place beyond the rounding place. Since you are rounding correct to hundredths (2 places), you must have 3 decimal places to the right of the decimal point:

```
      4.266
 18 ) 76.800
      72
      ----
       4 8
       3 6
       ----
       1 20
       1 08
       ----
         12
```

Add 2 zeros to carry out the division to 3 places.

Now, round the quotient correct to hundredths:

$$4.266 = 4.27$$

THE DECIMAL DIVISOR

When your divisor has a decimal point, follow these steps.

Step 1: Move the decimal point in the divisor as many places to the right as needed to make it a whole number.

Step 2: Move the decimal point in the dividend the **same number** of places as you did in the divisor, attaching zeros if necessary.

Step 3: Place the decimal point in the quotient directly above the newly located decimal point in the dividend.

Step 4: Divide.

EXAMPLE 1: Divide 1.44 by .12

Solution:

$$.12.\overline{)1.44.}$$

Move the decimal point 2 places to the right in the divisor, and 2 places to the right in the dividend.

Move the newly located decimal point into the quotient, and divide:

```
       12.
 12 ) 1 44.
      1 2
      ----
        24
        24
        --
         0
```

The decimal point is placed above its newly located place in the dividend.

EXAMPLE 2: Divide 135 by 5.7 and round the quotient correct to the nearest tenth.

Solution:

$$5.7.\overline{)135.0.}$$

Move the decimal point one place to the right in the divisor.

Move the decimal point one place to the right in the dividend. Since the dividend is a whole number, you must add a zero.

Now, divide as usual (remember, we are rounding the quotient to the nearest tenth):

```
       23.68
 57 ⟌ 1350.00
      114
      ----
       210
       171
       ----
        39 0
        34 2
        ----
         4 80
         4 56
         ----
           24
```

Remember: to round to the nearest tenth (one place to the right of the decimal point), you must carry the division out to 2 places and round back to one.

Answer: 23.68 = 23.7

EXAMPLE 3: Divide 675 by 3.28 and round the quotient to the nearest thousandth.

Solution:

```
           205.7926
 3.28. ⟌ 67500.0000
         656
         -----
          1900
          1640
          -----
           260 0
           229 6
           -----
            30 40
            29 52
            -----
               880
               656
               ----
               2240
               1968
               ----
                272
```

To round to thousandths (3 places) carry the division out to the fourth place and round to the third place.

Answer: 205.7926 = 205.793

Work Space

Name ______________________________ Date ______________________________

UNIT 36 Exercises

Divide each of these problems. Where needed, round off to the nearest thousandth (3 places).

1. $8\overline{).248}$
2. $7\overline{).286}$
3. $5\overline{)2.45}$
4. $12\overline{)4.88}$
5. $9\overline{)24.5}$
6. $3\overline{).033}$
7. $8\overline{)2.49}$
8. $2\overline{).0012}$
9. $5\overline{).028}$

ANSWERS

1. ____________
2. ____________
3. ____________
4. ____________
5. ____________
6. ____________
7. ____________
8. ____________
9. ____________

10. $7\overline{)34.765}$

11. $15\overline{).453}$

12. $21\overline{).4221}$

ANSWERS

10. ____________
11. ____________
12. ____________
13. ____________
14. ____________
15. ____________
16. ____________
17. ____________
18. ____________

Divide each of these problems. Give answers in dollar amounts rounded to the nearest cent (nearest hundredth).

13. $112.50 ÷ 15

14. $1,245.18 ÷ 12

15. $3,456.82 ÷ 25

16. $5,693.57 ÷ 8

17. $2,458.24 ÷ 14

18. $12,345.15 ÷ 212

Name ______________________ Date ______________________

19. $24,568.12 ÷ 240

20. $32,618.40 ÷ 30

Divide each of these problems. Where necessary, round answers to the nearest hundredth (two places).

21. $.02\overline{).45}$

22. $.08\overline{)1.25}$

23. $.5\overline{)25}$

24. $1.2\overline{)12.78}$

25. $.002\overline{)8}$

26. $.045\overline{)4.5}$

27. $.9\overline{)81}$

28. $.08\overline{)6.4}$

29. $.007\overline{).49}$

30. $.25\overline{)26.75}$

31. $.09\overline{)27.29}$

32. $3.5\overline{)142.3}$

ANSWERS

19. ______
20. ______
21. ______
22. ______
23. ______
24. ______
25. ______
26. ______
27. ______
28. ______
29. ______
30. ______
31. ______
32. ______

33. $.045\overline{)345.6}$

34. $.40\overline{)5.82}$

35. $1.4\overline{)50}$

ANSWERS

33. __________
34. __________
35. __________
36. __________
37. __________
38. __________
39. __________
40. __________
41. __________
42. __________
43. __________
44. __________

36. $.23\overline{)428}$

37. $.08\overline{)30.2}$

38. $5.25\overline{)75}$

39. $.0002\overline{)1.2}$

40. $.5\overline{)235.6}$

41. $.075\overline{)12.4}$

42. $.51\overline{)221}$

43. $2.4\overline{)3.6}$

44. $.03\overline{)30}$

Name ______________________ Date ______________________

Divide these problems and round off to the nearest cent.

ANSWERS

45. $25\overline{)\$525.95}$

46. $40\overline{)\$324.58}$

47. $30\overline{)\$95.80}$

48. $36\overline{)\$568.20}$

49. $52\overline{)\$454.98}$

50. $25\overline{)\$125.49}$

45. ____________
46. ____________
47. ____________
48. ____________
49. ____________
50. ____________

Work Space

Name ______________________ Date ______________________

UNIT 36 Applications

Where appropriate, round all answers to even cents (even hundredths).

ANSWERS

1. ____________
2. ____________
3. ____________

1. The total cost of a new car, including finance charges, is $12,456.88. If this amount can be paid in 48 monthly installments, what is the amount of each payment?

2. The Brewer Wholesale Company had 500 product catalogs printed at a total cost of $26,250. What is the cost of each catalog?

3. Professor Samuel Moore drove 1,456.8 miles on a recent lecture tour. How many miles per gallon did his car average if he bought a total of 56 gallons of gas?

4. Mendocin Company paid $945.52 to have a conference room carpeted. What was the cost per square yard if the area of the conference room is 50 square yards?

ANSWERS

4. ____________
5. ____________
6. ____________
7. ____________

5. What is the cost of an orange if a dozen cost $1.89?

6. Otto paid $13.86 to have his gas tank filled. How many gallons of gas did he purchase if one gallon costs $.99?

7. Franklin Hobbs earned $220 last week. If he earns $5.50 an hour, how many hours did he work?

Name ______________________ Date ______________________

8. As part of a profit sharing plan, Video Products, Inc. is going to distribute $12,450.80 to its 300 employees. If each employee is to get an equal amount, what is the share of each employee?

ANSWERS

8. ____________

9. ____________

10. ____________

9. Margaret Komendantov bought a stereo for $582 on time. How many payments will she make if one payment is $48.50?

10. If a gross (144) of writing pads costs $129.60, what is the cost per pad?

Work Space

UNIT 37

Dividing Decimals Using Shortcuts

In Unit 34, you worked with some shortcuts that can help you when multiplying decimal numbers. In this unit, we are going to look at some common shortcuts that help you when dividing decimal numbers.

DIVIDING BY A POWER OF 10

To divide a decimal number by a power of 10 (10, 100, 1,000, etc.), move the decimal point in the dividend **to the left** the same number of places as there are zeros in the divisor.

EXAMPLE 1: Divide: 645.8 ÷ 10
Solution:
Move the decimal point one place to the left:

$$645.8 \div 10 = 645.8 = 64.58$$

EXAMPLE 2: Divide: 56.45 ÷ 1,000
Solution:
Move the decimal point 3 places to the left:

$$56.45 \div 1{,}000 = 056.45 = .05645$$

Note: a zero was placed to the right of the decimal as a placeholder.

EXAMPLE 3: Divide: 324 ÷ 100
Solution:
Move the decimal point two places to the left:

$$324 \div 100 = 324. = 3.24$$

The decimal point in a whole number is always after the right-hand digit.

DIVIDING BY 25 AND 50

To divide a number by 25, first multiply by 4, then divide by 100 (move decimal point two places to the left). For example, to divide 346.8 by 25:

Multiply by 4:

$$\begin{array}{r} 346.8 \\ \times \quad 4 \\ \hline 1387.2 \end{array}$$

Divide by 100:

$$1387.2 = \mathbf{13.872}$$

To divide a number by 50, first multiply by 2, then divide by 100. For example, to divide 468.42 by 50:

Multiply by 2:

$$\begin{array}{r} 468.42 \\ \times \quad 2 \\ \hline 936.84 \end{array}$$

Divide by 100:

$$936.84 = \mathbf{9.3684}$$

Remember.

1. To *divide* a number by a power of 10, move the decimal point in the dividend to the **left** the same number of places as there are zeros in the divisor. Be careful; if you move the decimal point to the **right,** you will be multiplying, not dividing.
2. To divide a number by 25, first multiply by 4, and then point off 2 places in the product.
3. To divide a number by 50, first multiply by 2, and then point off 2 places in the product.

Name ______________________ Date ______________________

UNIT 37 Exercises

Divide each of the following problems in your head.

1. 345.8 ÷ 10 2. 48.2 ÷ 100 3. 2.7 ÷ 100

4. 124.89 ÷ 10 5. 1.2 ÷ 1,000 6. 15.45 ÷ 1,000

7. 2,458.2 ÷ 1,000 8. 3.48 ÷ 100 9. .07 ÷ 100

10. 4.64 ÷ 1,000 11. 500 ÷ 10,000 12. 89.1 ÷ 100

13. 3,456.89 ÷ 10,000 14. 318.009 ÷ 10 15. 8 ÷ 1,000

Divide each of these numbers by 10 in your head.

16. 80 17. 565 18. 3.45 19. 2.345

20. 565.8 21. 4,692 22. .4095 23. 748.775

Divide each of these numbers by 100 in your head.

24. 500 25. 658 26. 45.8 27. 238.65

28. .855 29. 12.5 30. 17.459 31. 675.4

ANSWERS

1. ________
2. ________
3. ________
4. ________
5. ________
6. ________
7. ________
8. ________
9. ________
10. ________
11. ________
12. ________
13. ________
14. ________
15. ________
16. ________
17. ________
18. ________
19. ________
20. ________
21. ________
22. ________
23. ________
24. ________
25. ________
26. ________
27. ________
28. ________
29. ________
30. ________
31. ________

Divide each of these numbers by 1,000 in your head.

32. 3,000 33. 33,500 34. 289.5 35. .8

36. 3,455.78 37. 401.89 38. 543.21 39. .055

Divide each of these numbers by 10,000 in your head.

40. 5,689 41. 4,000 42. 40,000

43. 12,345.89 44. 567.8 45. 178,563.76

46. 211.91 47. 3,456.72

Using shortcuts, divide each of these numbers by 25.

48. 400 49. 32.4 50. 12.82

51. 445.24 52. 1,632.44 53. 1,232.36

54. 6,416.48 55. 3,200

Using shortcuts, divide each of these numbers by 50.

56. 2,000 57. 4,200 58. 3,820.22

ANSWERS

32. ____________
33. ____________
34. ____________
35. ____________
36. ____________
37. ____________
38. ____________
39. ____________
40. ____________
41. ____________
42. ____________
43. ____________
44. ____________
45. ____________
46. ____________
47. ____________
48. ____________
49. ____________
50. ____________
51. ____________
52. ____________
53. ____________
54. ____________
55. ____________
56. ____________
57. ____________
58. ____________

Name ______________________ Date ______________________

59. 240.84

60. 4,640.32

61. 2.008

62. 6,346.82

63. 12.78

ANSWERS

59. ____________

60. ____________

61. ____________

62. ____________

63. ____________

Work Space

Name ______________________ Date ______________________

UNIT 37 Applications

1. If 100 pens cost $56, what is the cost of one pen?

2. Steve Campione earned $87.50 in 10 hours. What is his wage per hour?

3. The Kissel Company's annual picnic and softball game cost $1,285. If this cost is to be shared equally by 100 employees, what is the cost per employee?

4. West Concrete Company paid $12,485 for 10,000 advertising leaflets. To the nearest cent, what is the cost per leaflet?

5. The guests at a pool party sent out for pizza. The total cost of the pizza was $65, which was shared equally by 25 people. What is the cost per person?

6. If a box of thumb tacks that has 100 tacks costs $3.50, what is the cost of one tack?

7. Richard Marks paid $125 for snacks and gifts for a party for his fourth-grade class. What was the cost per child if his class has a total of 50 students?

ANSWERS

1. ____________
2. ____________
3. ____________
4. ____________
5. ____________
6. ____________
7. ____________

8. If 1,000 pounds of coffee has a wholesale price of $2,950, what is the cost of one pound?

9. On their last vacation trip, the Vella family traveled a total of 1,000 miles. What was their cost per mile if they spent $120 for gas?

10. Southland Foods had sales of $144,318 over a 100-day period. What was the average daily sales?

ANSWERS

8. ____________

9. ____________

10. ____________

UNIT 38 Equivalent Decimal and Common Fractions

Remember that a decimal is a special kind of fraction. Thus, all decimal fractions can be changed to equivalent common fractions; and all common fractions can be changed to equivalent decimal fractions. In business, common fractions are often changed to their decimal fraction equivalent to make problems easier to work with. Let's look at how this is done.

CHANGING A DECIMAL FRACTION TO AN EQUIVALENT COMMON FRACTION

An easy way to change a decimal fraction to a common fraction is to follow these steps:

Step 1: Read the decimal fraction aloud.

Step 2: Write the decimal as a common fraction.

Step 3: Reduce the common fraction to lowest terms, if needed.

EXAMPLE 1: Change .125 to a common fraction in lowest terms.

Solution:

Step 1: Read the decimal fraction: "one hundred twenty-five thousandths"

Step 2: Write the decimal as a common fraction:

$$\frac{125}{1{,}000}$$

Step 3: Reduce the common fraction to lowest terms:

$$\frac{125}{1{,}000} = \frac{1}{8}$$

EXAMPLE 2: Change .002 to a common fraction in lowest terms.

Solution:

$$.002 = \frac{2}{1{,}000} = \frac{1}{500}$$

EXAMPLE 3: Change 5.25 to a common fraction in lowest terms.

Solution:

$$5.25 = 5\frac{25}{100} = 5\frac{1}{4}$$

Another way to change a decimal to a fraction is to count the number of places after the decimal point. Then, for your denominator, write the number 1 followed by as many zeros as there are places after the decimal point. For example:

$$.4 = \frac{4}{10}$$

One place after the decimal point = a fraction with one zero in the denominator.

$$.28 = \frac{28}{100}$$

Two places after the decimal point = a fraction with two zeros in the denominator.

$$.375 = \frac{375}{1{,}000}$$

Three places after the decimal point = a fraction with three zeros in the denominator.

CHANGING A COMMON FRACTION TO AN EQUIVALENT DECIMAL FRACTION

To change a common fraction to an equivalent decimal fraction, you divide the denominator into the numerator. You may have to add zeros to the numerator to get the division to come out evenly.

EXAMPLE 1: Change $\frac{5}{8}$ to an equivalent decimal fraction.
Solution:

$\frac{5}{8} = 8\overline{)5.000}$ = .625

4 8
20
16
40
40

We added three zeros to make the division come out evenly.

EXAMPLE 2: Change $\frac{1}{2}$ to an equivalent decimal fraction.
Solution:

$\frac{1}{2} = 2\overline{)1.0}$ = .5

1 0

We added one zero to make the division come out evenly.

For numbers that cannot be divided evenly, you normally carry the division correct to three places. For example, when you change $\frac{2}{3}$ to a decimal, you will keep getting a remainder no matter how many zeros you add:

$\frac{2}{3} = 3\overline{)2.000}$ = .666...

1 8
20
18
20
18
2

The three dots mean that the decimal will keep repeating, no matter how many zeros you add. It is read as **point 6 repeating.**

DECIMALS AND FRACTIONS IN A SINGLE PROBLEM

If you are working with a problem that contains both decimal fractions and common fractions, you must change the problem so that all fractions will be stated as either common fractions or decimal fractions.

EXAMPLE 1: Add: $112.5 + 15\frac{3}{4} + 10\frac{1}{8}$
Solution:
As a common fraction

Change the decimal fraction to a common fraction.

$$\begin{array}{lcl} 112.5 & & 112\tfrac{1}{2} = 112\tfrac{4}{8} \\ 15\tfrac{3}{4} & = & 15\tfrac{3}{4} = 15\tfrac{6}{8} \\ 10\tfrac{1}{8} & & 10\tfrac{1}{8} = 10\tfrac{1}{8} \\ & & 137\tfrac{11}{8} = 138\tfrac{3}{8} \end{array}$$

As a decimal fraction

Change the common fraction to a decimal fraction.

$$\begin{array}{lcr} 112.5 & & 112.500 \\ 15\tfrac{3}{4} & = & 15.750 \\ 10\tfrac{1}{8} & & 10.125 \\ & & 138.375 \end{array}$$

EXAMPLE 2: Multiply: $3.25 \times 2\frac{1}{2}$
Solution:
As a common fraction

$$3\tfrac{1}{4} \times 2\tfrac{1}{2} = \tfrac{13}{4} \times \tfrac{5}{2} = \tfrac{65}{8} = 8\tfrac{1}{8}$$

As a decimal fraction

$$\begin{array}{r} 3.25 \\ \times\ 2.5 \\ \hline 1625 \\ 650 \\ \hline 8125 = 8.125 \end{array}$$

Name ______________________________ Date ______________________________

UNIT 38 Exercises

Change these decimal fractions to common fractions in lowest terms.

1. .12 =
2. .015 =
3. .08 =
4. .002 =
5. .375 =
6. .025 =
7. .0025 =
8. .3 =
9. .03 =
10. .0075 =
11. .60 =
12. .425 =
13. .1245 =
14. .90 =
15. .00125 =
16. .45 =
17. .005 =
18. .0085 =
19. .625 =
20. .2 =
21. .09 =

ANSWERS

1. ____________
2. ____________
3. ____________
4. ____________
5. ____________
6. ____________
7. ____________
8. ____________
9. ____________
10. ____________
11. ____________
12. ____________
13. ____________
14. ____________
15. ____________
16. ____________
17. ____________
18. ____________
19. ____________
20. ____________
21. ____________

Change these common fractions to equivalent decimal fractions. Round off correct to the third place where applicable.

22. $\frac{7}{8}$ 23. $\frac{3}{4}$ 24. $\frac{1}{3}$

25. $\frac{4}{5}$ 26. $\frac{5}{7}$ 27. $\frac{2}{3}$

28. $\frac{1}{7}$ 29. $\frac{11}{12}$ 30. $\frac{1}{20}$

31. $\frac{5}{6}$ 32. $\frac{1}{9}$ 33. $\frac{1}{8}$

34. $\frac{1}{6}$ 35. $\frac{4}{9}$ 36. $\frac{7}{11}$

37. $\frac{1}{2}$ 38. $\frac{6}{7}$ 39. $\frac{5}{12}$

40. $\frac{17}{18}$ 41. $\frac{15}{19}$

ANSWERS

22. ________
23. ________
24. ________
25. ________
26. ________
27. ________
28. ________
29. ________
30. ________
31. ________
32. ________
33. ________
34. ________
35. ________
36. ________
37. ________
38. ________
39. ________
40. ________
41. ________

Name ______________________ Date ______________________

Give the equivalent fraction and decimal answers to each of the following. Round answers to three places where applicable.

	Fractional Method	Decimal Method
42. $12\frac{1}{8} + 8\frac{3}{4}$	________	________
43. $125.55 + 2\frac{3}{8} + 1\frac{4}{5}$	________	________
44. $312\frac{7}{8} + 225.75$	________	________

	Fractional Method	Decimal Method
45. $29.25 - 3\frac{4}{5}$	__________	__________
46. $315\frac{1}{8} - 15.4$	__________	__________
47. $3\frac{4}{5} \times 2.8$	__________	__________
48. $12.2 \times 10\frac{3}{5}$	__________	__________

Name ______________________ Date ______________________

	Fractional Method	Decimal Method
49. $16.4 \div 4\frac{2}{5}$	________	________
50. $40\frac{5}{8} \div 2\frac{1}{10}$	________	________

Work Space

UNIT 39 Understanding Percents

In earlier units we studied two ways of expressing a part of a whole unit—decimals and fractions. If, for example, you want to express "one part out of a total of four parts," the fraction $\frac{1}{4}$ will do this. The decimal .25 (twenty-five hundredths) can also be used to express one part of a total of four parts. Thus, $\frac{1}{4}$ and .25 are different ways of expressing the same value.

A third way to express a part of a whole unit is through **percent,** which means *by the hundred* or **out of a hundred.** Thus, **one percent** is "one part out of a total of 100 parts," and **10 percent** is "10 parts out of a total of 100 parts." A dollar, for example, has 100 cents; thus, 10 cents equals 10 percent of a dollar.

A PERCENT IS A FRACTION WITH A DENOMINATOR OF 100

A percent is like a decimal in that they are both fractions with an understood denominator. The denominator of a decimal fraction, as you recall, is always 10, 100, 1,000, or some other power of 10; the denominator of a percent is always 100.

When working with percents, we often use the percent sign (%) in place of the word "percent." Thus, 15% is read as "15 percent," and $12\frac{1}{2}\%$ is read as "twelve and one-half percent."

Because decimals, percents, and fractions are all different ways of expressing the same values, we can easily change one form of expression to another. We will see how that is done next.

CHANGING DECIMALS TO PERCENTS

To change a decimal to a percent, you move the decimal point two places to the right and add on the percent sign.

EXAMPLES: Change the following decimals to percents:

.12

.08

.375

Solution:

.12 = .12 = 12%

.08 = .08 = 8%

.375 = .375 = 37.5% (or $37\frac{1}{2}\%$)

Whole numbers and mixed numbers are changed to percents in the same way:

5 = 5.00 = 500%

2 = 2.00 = 200%

3.5 = 3.50 = 350%

CHANGING PERCENTS TO DECIMALS

To change a percent to a decimal, you reverse the above operation. That is, you move the decimal point two places to the left and drop the percent sign.

EXAMPLES: Change the following percents to decimals:

18%

6%

67.5%

2.5%

Solution:

18% = .18

06% = .06

67.5% = .675

02.5% = .025

CHANGING FRACTIONS TO PERCENTS

To change a fraction to a percent, first change the fraction to an equivalent decimal; then, move the decimal point two places to the right and add on the percent sign.

EXAMPLES: Change the following fractions to percents:

$\frac{1}{2}$

$\frac{3}{4}$

$2\frac{1}{4}$

Solution:
Change the fractions to decimals, then change the decimals to percents.

Fraction ⟶	Decimals ⟶	Percents
$\frac{1}{2} = 1 \div 2 = \rightarrow$	$.5 = .50 = \rightarrow$	50%
$\frac{3}{4} = 3 \div 4 = \rightarrow$	$.75 = .75 = \rightarrow$	75%
$2\frac{1}{4} = 9 \div 4 = \rightarrow$	$2.25 = 2.25 = \rightarrow$	225%

Remember: $2\frac{1}{4} = \frac{(4 \times 2 = 8) + 1}{4} = \frac{9}{4}$

Look at the last example; another way we could have changed $2\frac{1}{4}$ to 225% is as follows:

$$2\frac{1}{4} = 2.25 = 2.25 = 225\%$$

CHANGING PERCENTS TO FRACTIONS

To change a percent to a fraction, first change the percent to a decimal; then, change the decimal to a fraction. Reduce the fraction to lowest terms, if needed.

EXAMPLES: Change the following percents to fractions:

4%

25%

80%

2.75%

Solution:
Change the percent to a decimal, then change the decimal to a fraction.

Percent ⟶	Decimal ⟶	Fraction
$4\% = .04\% = \rightarrow$	$.04 = \rightarrow$	$\frac{4}{100} = \frac{1}{25}$
$25\% = .25\% = \rightarrow$	$.25 = \rightarrow$	$\frac{25}{100} = \frac{1}{4}$
$80\% = .80\% = \rightarrow$	$.80 = \rightarrow$	$\frac{80}{100} = \frac{4}{5}$
$2.75\% = .0275\% = \rightarrow$	$.0275 = \rightarrow$	$\frac{275}{10{,}000} = \frac{11}{400}$

Another way to change a percent to a fraction is to remember that a percent is a fraction with a denominator of 100:

$$5\% = \frac{5}{100} = \frac{1}{20}$$

$$40\% = \frac{40}{100} = \frac{2}{5}$$

$$55\% = \frac{55}{100} = \frac{11}{20}$$

Remember:

1. Percent means "by the hundred."
2. A percent is a special type of fraction that always has an understood denominator of 100:

$$50\% = \frac{50}{100}$$

$$6\% = \frac{6}{100}$$

$$33\% = \frac{33}{100}$$

3. Since fractions, decimals, and percents are different ways of expressing the same values, you can easily change from one form of expression to another.

Work Space

Name ______________________ Date ______________________

UNIT 39 Exercises

Fill in the following blanks.

1. $\frac{10}{100}$ = ______ % 2. $\frac{6}{100}$ = ______ % 3. $\frac{85}{100}$ = ______ %

4. $\frac{5}{100}$ = ______ % 5. $\frac{7}{100}$ = ______ % 6. $\frac{125}{100}$ = ______ %

7. $\frac{2}{100}$ = ______ % 8. $\frac{375}{100}$ = ______ % 9. $\frac{72.5}{100}$ = ______ %

10. 45 out of 100 = ______ % 11. 31 out of 100 = ______ %

12. $2\frac{1}{2}$ out of 100 = ______ % 13. 99 out of 100 = ______ %

14. $12\frac{1}{3}$ out of 100 = ______ % 15. 1 out of 100 = ______ %

16. 4 hundredths = ______ % 17. 35 hundredths = ______ %

18. 225 hundredths = ______ % 19. 17 hundredths = ______ %

20. 12.5 hundredths = ______ % 21. 100 hundredths = ______ %

Change each of these decimals to percents.

22. .45 = 23. .98 = 24. 1.05 = 25. 2.75 =

26. .375 = 27. .17 = 28. 1.24 = 29. .02 =

30. 3.50 = 31. 5 = 32. .376 = 33. .50 =

ANSWERS

22. ______
23. ______
24. ______
25. ______
26. ______
27. ______
28. ______
29. ______
30. ______
31. ______
32. ______
33. ______

Change each of these percents to decimals.

34. 5% =

35. 8% =

36. 12% =

37. 125% =

38. 28.5% =

39. 200% =

40. 34.8% =

41. 2.5% =

42. 100% =

Change each of these fractions to percents.

43. $\frac{3}{4}$ =

44. $\frac{1}{2}$ =

45. $\frac{4}{5}$ =

46. $\frac{3}{5}$ =

47. $2\frac{1}{2}$ =

48. $\frac{1}{4}$ =

49. $4\frac{3}{4}$ =

50. $\frac{5}{6}$ =

51. $\frac{1}{8}$ =

ANSWERS

34. __________
35. __________
36. __________
37. __________
38. __________
39. __________
40. __________
41. __________
42. __________
43. __________
44. __________
45. __________
46. __________
47. __________
48. __________
49. __________
50. __________
51. __________

Name ______________________ Date ______________________

Change each of these percents to fractions. Reduce to lowest terms, where needed.

52. 24% =

53. 45% =

54. 18% =

55. 15% =

56. $2\frac{1}{2}\%$ =

57. $6\frac{1}{4}\%$ =

58. 225% =

59. 5.8% =

60. .5% =

ANSWERS

52. ______
53. ______
54. ______
55. ______
56. ______
57. ______
58. ______
59. ______
60. ______

Complete the following.

	Fraction	Decimal	Percent
62.	$\frac{4}{5}$	______	______
63.	______	.80	______
64.	______	______	10%
65.	$\frac{2}{3}$	______	______
66.	______	.025	______
67.	______	______	125%
68.	______	.95	______
69.	$\frac{9}{10}$	______	______

Work Space

Name ____________________ Date ____________________

UNIT 39 Applications

1. 52% of the workers at MUTECH Company are women. For each 100 workers, how many are women?

2. One Monday morning, 20% of the students at Eastdale High were absent. What percent of the student body was present?

3. If a suit is marked down by 30%, what percent of the selling price will be paid?

4. Glenn Seymour received a 25% raise in pay. What percent is his new salary compared to his old salary?

5. If 71% of the employees at Denver Company have health insurance, what percent do not?

6. In a recent election, there was a 90% voter turnout. What percent of voters did not vote?

ANSWERS

1. ________
2. ________
3. ________
4. ________
5. ________
6. ________

7. Jane McKay is in the 40% tax bracket. What percent of her pay does she take home?

8. What percent of a dollar is 75 cents?

9. In a certain Florida town, 8 out of 10 people are retired. Express this as a percent.

10. What is 25% of a dollar?

ANSWERS

7. ____________

8. ____________

9. ____________

10. ____________

UNIT

40 Working with the Base, Rate, and Percentage

To start our discussion, let's look at a transaction that is typical of the many thousands of business transactions that occur each day in our country.

> John Dorman is a real estate salesperson. Last week John sold a house for $90,000. His 6% commission for selling the house was $5,400.

This application of percent identifies three elements that you should be very familiar with: the base, rate, and percentage.

> **Base.** The **base** is the whole value, or the beginning quantity. In the above example, the base is the price of the house, $90,000. The word "of" or "part of" is often written before the base. Thus, writing the above problem as "6% **of** $90,000 equals the commission on the sale" would help identify the $90,000 as the base in this problem.
>
> **Rate.** The **rate** is a percent (such as 8%, 10%, $12\frac{1}{2}\%$, and so on). The rate indicates the part of the base that is to be calculated. In the above example, the rate is 6%. As a general rule, the rate can be easily identified because it is usually written with the percent sign.
>
> **Percentage.** The **percentage** is the result of multiplying the base by the rate. In the above example, the percentage is $5,400, which resulted from multiplying the base, $90,000, by the rate, 6% ($90,000 × 6% = $5,400).

> *Caution:* Don't confuse the "percent" and the "percentage." The percent is a rate; the percentage is the product of the base times the rate.

We have now identified the three elements of a percent problem: the base, rate, and percentage. If two of these elements are known, the third can always be calculated. Let's look at how to find a missing element when the other two are given.

FINDING THE PERCENTAGE (P)

Since the percentage is the product of the base and the rate, we can use this formula to find the percentage:

$$\text{Base} \times \text{Rate} = \text{Percentage}$$

or

$$B \times R = P$$

EXAMPLE 1: If the base is 800 and the rate is 8%, find the percentage.
Solution:

```
     800 ←——— base
  ×  .08 ←——— rate
    6400 = 64 ←——— percentage
```

> *Caution*: Unless you are working with a calculator with a "%" key, you must change the percent to an equivalent decimal in order to work with it.

EXAMPLE 2: If the base is \$400 and the rate is $2\frac{1}{2}\%$, find the percentage.

```
   $400 <------ base
×  .025 <------ rate
   2000
   800
  10000 = $10 <------ percentage
```

EXAMPLE 3: Find 25% of 600.
Solution:

$B \times R = P$

$600 \times .25 = 150$

> Helpful hint: when a problem is written like this, the word "of" means to multiply.

In some problems, you will need to find the percentage even though the problem does not directly say so.

EXAMPLE: Jim Pearson earns \$900 a month. He spends 20% of this amount for rent. Determine his rent expense.
Solution:

Base = \$900 (remember, the base is the whole quantity)

Rate = 20% (remember, the "%" mark will usually identify the rate)

Percentage = ?

$$B \times R = P$$

$$\$900 \times .20 = \$180$$

When solving percent problems, it often helps to use a pie diagram, like this:

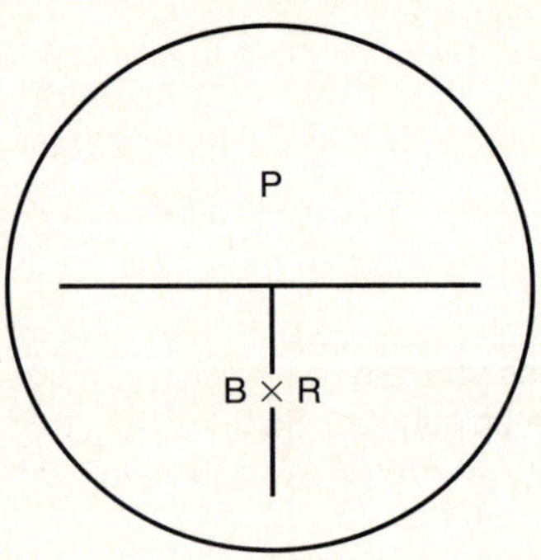

To use the pie diagram, place your finger over the missing element; the elements shown will then show you how to find the missing element. To find the percentage, for example, place a finger over the P:

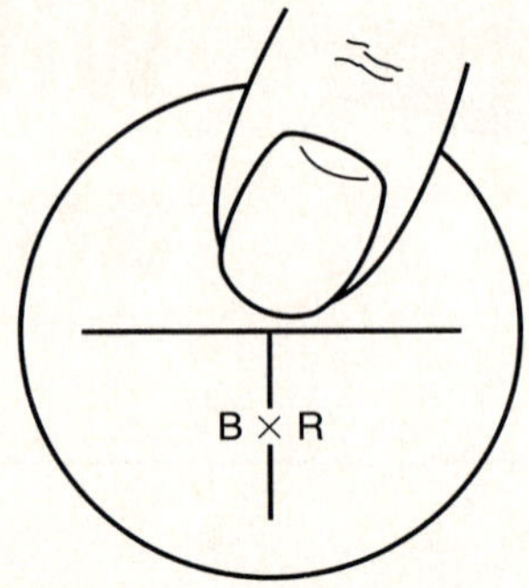

= B × R = P

EXAMPLE: Using a pie diagram, find the percentage when the base is 500 and the rate is 12%.
Solution:

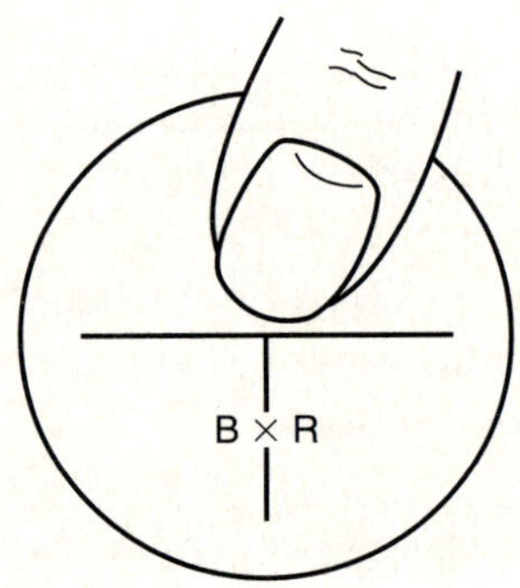

B *R* *P*
= 500 × .12 = 60

FINDING THE RATE (R)

The rate can be found as follows:

$$P \div B = R$$

EXAMPLE: The percentage is \$54 and the base is 900. What is the rate?
Solution:

$P \div B = R$

$\$54 \div \$900 = 6\%$

```
       .06  = 6%
900 ) 54.00
      54 00
```

The use of a pie diagram will also help when looking for the rate. For example, let's solve the preceding example with the use of a pie diagram.

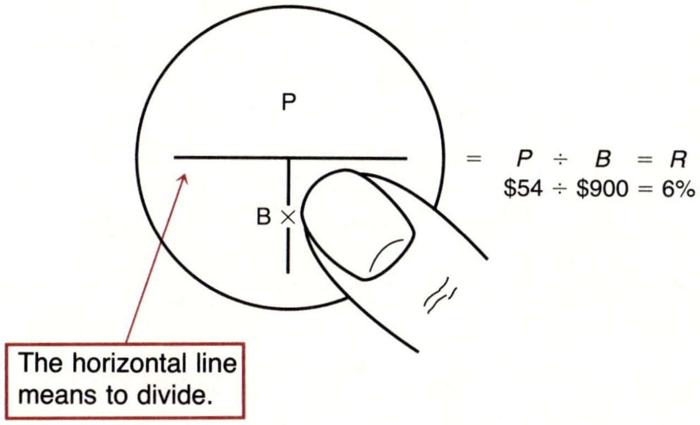

Let's look at another example.

EXAMPLE: Dean Watson received a $32.50 commission for selling $650 worth of merchandise. What was his rate of commission?
Solution:

B = $650 (remember, the base is the beginning quantity)

P = $32.50 (the part of the whole)

R = ?

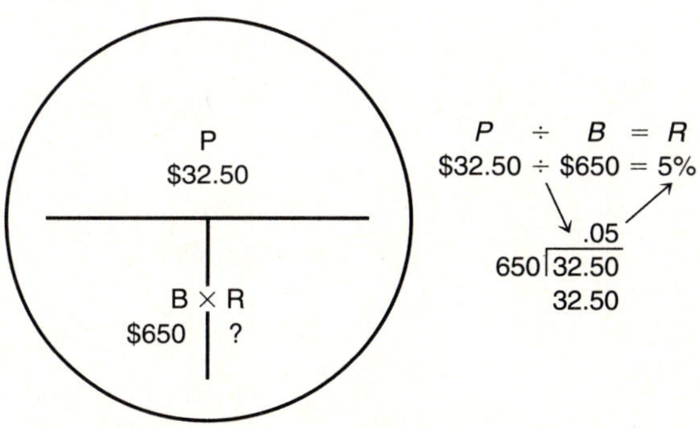

Check: $650 × .05 = $32.50

FINDING THE BASE

The base can be found as follows:

$$P \div R = B$$

EXAMPLE: The percentage is $36 and the rate is 9%. Find the base.
Solution:

P	$\div$	R	$=$	B
$36	÷	.09	=	$400

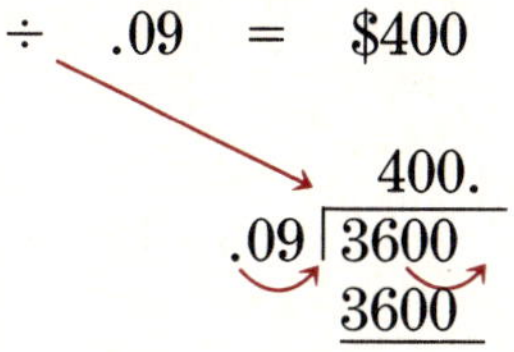

Check: $400 × .09 = $36

Again, we can use a pie diagram to help find the missing element—the base, in this case. Let's look again at the preceding example.

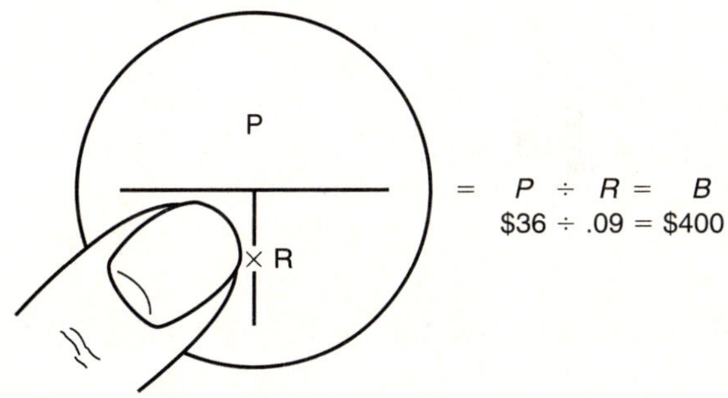

Let's look at another example where we need to calculate the base.

EXAMPLE: 72 is 8% of what number?
Solution:

P = 72 (the part of the whole)

R = 8% (the "%" sign identifies the rate)

B = ? (the beginning whole quantity is missing)

P	$\div$	R	$=$	B
72	÷	.08	=	900

Check: 900 × .08 = 72

Remember.

$$B \times R = P$$

$$P \div B = R$$

$$P \div R = B$$

Work Space

Name ______________________ Date ______________________

UNIT 40 Exercises

Find the percentage in each of these problems. Remember to change the rates to decimals before you multiply.

1. 5% of 48
2. 20% of 24
3. 75% of 100
4. 8% of 500
5. 90% of 2,000
6. 42% of 300
7. 10% of 50
8. 16% of 250
9. 11% of 400
10. 60% of $3,500
11. 55% of $350
12. 37% of $1,200
13. 14% of $525
14. 28% of $960

ANSWERS

1. ______
2. ______
3. ______
4. ______
5. ______
6. ______
7. ______
8. ______
9. ______
10. ______
11. ______
12. ______
13. ______
14. ______

15. 25% of $1,650

16. 95% of $585

ANSWERS

15. ____________

16. ____________

17. ____________

18. ____________

17. 6% of $35,000

18. 9% of $45,600

Find the rate in each of these problems.

Base	Percentage	Rate
19. $ 6,000	$ 600	______
20. 8,500	1,020	______
21. 2,450	490	______
22. 9,000	450	______
23. 12,250	9,800	______

Name ______________________ Date ______________________

	Base	Percentage	Rate
24.	18,625	5,960	_____
25.	4,000	2,000	_____
26.	12,000	240	_____
27.	1,500	105	_____

Find the base in each of these problems.

	P	R	B
28.	300	10%	_____
29.	225	8%	_____
30.	630	12%	_____

P	R	B
31. 4,500	9%	______
32. 3,600	6%	______
33. 900	3%	______
34. 450	5%	______
35. 1,800	9%	______
36. 7,200	8%	______

Name ____________________ Date ____________________

Find the missing element in each of these problems.

	Base	Rate	Percentage
37.	$30,000.00	15%	________
38.	$ 6,800.00	___	$ 272.00
39.	________	40%	$6,000.00
40.	$ 455.80	8%	________
41.	________	6%	$ 735.30
42.	$12,890.75	___	$1,289.08

Work Space

Name ______________________ Date ______________________

	Base	Percentage	Rate
24.	18,625	5,960	____
25.	4,000	2,000	____
26.	12,000	240	____
27.	1,500	105	____

Find the base in each of these problems.

	P	R	B
28.	300	10%	____
29.	225	8%	____
30.	630	12%	____

P	R	B
31. 4,500	9%	____
32. 3,600	6%	____
33. 900	3%	____
34. 450	5%	____
35. 1,800	9%	____
36. 7,200	8%	____

Name ____________________ Date ____________________

UNIT 40 Applications

1. Identify the base, rate, and percentage in each of the following statements:

   a. 25% of 300 is 75

   b. $200 is 20% of $1,000

   c. $6,000 × .08 = $480

2. What word, or words, normally indicate the base in a problem?

3. Jane Murrie is a sales representative for a large grocery company. Jane is paid a 6% commission on all goods she sells. What is her commission in a month in which her total sales amounted to $55,250?

4. Jim McGuire pays 30% of his earnings in taxes. What are his taxes in a week in which he earned $350?

5. On his last payday, Robert Santaliz deposited $60 in his savings account. If the deposit was 5% of his total earnings, what were his total earnings?

6. Reda Kinza, a real estate salesperson, sold a house for $75,000. What is the amount of her commission if her rate of commission is 7%?

ANSWERS

1. a. ____________
   ____
   ____
   b. ____________
   ____
   ____
   c. ____________
   ____
   ____
2. ____________
3. ____________
4. ____________
5. ____________
6. ____________

7. Jeff Boling sold a car at a profit of $400. If his profit was 40% of what he paid for the car, how much did the car cost him?

ANSWERS

7. ____________

8. ____________

9. a. ____________

   b. ____________

10. ____________

11. ____________

12. ____________

8. At a local school, 250 students are taking a bookkeeping course. If this is 10% of the total school enrollment, how many students does the school have?

9. Jenna Rubin paid a 6% sales tax on the purchase of a $12,000 car. What is (a) the amount of sales tax and (b) the total cost of the car after sales taxes are added?

10. An employment test given by a telephone company was passed by 35% of the people who took the test. How many passed if a total of 500 people took the test?

11. Last year the operating expenses of SmartTech Company were 40% of the company's sales revenue. If the operating expenses were $360,000, what were the total sales?

12. After payroll taxes are withheld, the amount of Ray Kinney's weekly payroll check is $180. What rate of taxes is he paying if he earns $250 before taxes?

UNIT 41 Additional Applications of Percents

In the preceding two units, we reviewed percents and we studied their application to common business situations. In this unit, we are going to look at some additional topics in percent.

FRACTIONAL PARTS OF 1%

We have learned that 1% means "1 part out of 100 parts." In other words, 1% is a fraction of 100%. But sometimes percents are expressed as fractions of 1%. For example, $\frac{1}{4}\%$ means "$\frac{1}{4}$ of 1%" and $\frac{2}{3}\%$ means "$\frac{2}{3}$ of 1%." Such expressions are fractions of 1%, *not fractions of 100%*. So you should be careful when working with them.

When working with fractions of 1%, it is usually easier to change them to their decimal equivalents, which we do as follows:

To change a fraction of 1% to its decimal equivalent, you:

Step 1: Rewrite the common fraction as a decimal fraction.

Step 2: Drop the % sign and move the decimal point two places to the left.

EXAMPLE 1: Change $\frac{1}{4}\%$ to a decimal.
Solution:

Step 1: Rewrite the common fraction as a decimal:

$$\frac{1}{4}\% = .25\%$$

Since $\frac{1}{4} = .25$, $\frac{1}{4}\% = .25\%$

Step 2: Drop the % sign and move the decimal point two places to the left:

$$.25\% = 00.25\% = .0025$$

EXAMPLE 2: Change $\frac{3}{8}\%$ to a decimal.
Solution:

$$\frac{3}{8}\% = .375\% = .00.375\% = .00375$$

Since $\frac{3}{8} = .375$, $\frac{3}{8}\% = .375\%$

As we have learned, some fractions do not come out even when they are changed to decimal equivalents. When this happens, you must round off the decimal to a given place.

EXAMPLE : Change $\frac{2}{3}\%$ to a decimal rounded correct to 3 places.
Solution:

$$\frac{2}{3}\% = 3\overline{)2.0000}\ \ .6666 = .667\% = .00.667\% = .00667$$

Remember: to round a decimal correct to 3 places, you must carry out the decimal to 4 places and round back to 3.

FINDING THE PERCENT OF ONE NUMBER TO ANOTHER NUMBER

In business, you often have to find the percent of one number to another number. To do this, you start out by writing the numbers as a fraction. You then change the fraction to a percent.

EXAMPLE 1: 40 is what percent of 80?
Solution: Express 40 as a fraction of 80:

$$\frac{40}{80} = \frac{1}{2}$$

Change the fraction to a percent:

$$\frac{1}{2} = .50 = 50\%$$

Answer = 50%

Another way we could have solved this problem is by using the formula for finding the rate (studied in Unit 40):

$P \div B = R$

$40 \div 80 = 50\%$

$$80\overline{)40.00} = .50 = .50 = 50\%$$

EXAMPLE 2: Of the 200 workers at Arcon Industries, 25 were absent on Friday. Find the percent of those absent.

$$\frac{25}{200} = \frac{1}{8} = .125 = .125 = 12.5\% \text{ (or } 12\tfrac{1}{2}\%)$$

or

$\underline{P} \div \underline{B} = \underline{R}$

$25 \div 200 = 12.5\%$

Remember: the base is the beginning whole quantity.

The percentage is the part of the base.

FINDING THE PERCENT OF INCREASE OR DECREASE

Finding the percent of increase or decrease in amounts is a very common business application. For example, the manager of Lakeside Grocery noticed that produce sales rose from $16,000 in May to $20,000 in June. To compare the increase with other months, he calculated the percent of increase (or the rate of increase). To find the percent of increase (or decrease), you first find the amount of increase or decrease; you then divide the amount of increase or decrease by the original amount. We can state this in formula form, as follows:

$$\text{Percent of Increase or Decrease} = \frac{\text{Amount of Increase or Decrease}}{\text{Original Amount}}$$

EXAMPLE 1: If Lakeside Grocery's produce sales went from $16,000 in May to $20,000 in June, find the percent of increase.
Solution:
Find the amount of increase:

	$20,000 ←	new amount
−	16,000 ←	original amount
	$ 4,000 ←	amount of increase

Divide the amount of increase by the original amount:

$$\$16{,}000\overline{)\$4{,}000.00} = .25 = \textbf{25\% increase}$$

EXAMPLE 2: When MicroTech got a new computer system, the company was able to reduce its operating expenses from $80,000 in 19X1 to $72,800 in 19X2. What was the percent decrease?
Solution:

	$80,000 ←	original amount
−	72,800 ←	new amount
	$ 7,200 ←	amount of change

$$\$80{,}000\overline{)\$7{,}200.00} = .09 = \textbf{9\% decrease}$$

Name ______________________________ Date ______________________________

UNIT 41 Exercises

Change each of these to a decimal. Where necessary, round correct to three places.

1. $\frac{1}{2}\% =$

2. $\frac{5}{8}\% =$

3. $\frac{1}{10}\% =$

4. $\frac{3}{8}\% =$

5. $\frac{7}{10}\% =$

6. $\frac{4}{5}\% =$

7. $\frac{1}{5}\% =$

8. $\frac{7}{8}\% =$

9. $\frac{1}{3}\% =$

10. $\frac{3}{10}\% =$

11. $\frac{2}{5}\% =$

12. $\frac{2}{15}\% =$

13. $\frac{18}{25}\% =$

14. $\frac{9}{15}\% =$

ANSWERS

1. ____________
2. ____________
3. ____________
4. ____________
5. ____________
6. ____________
7. ____________
8. ____________
9. ____________
10. ____________
11. ____________
12. ____________
13. ____________
14. ____________

Complete the following. Where applicable round final answers correct to two places.

ANSWERS

15. ______
16. ______
17. ______
18. ______
19. ______
20. ______
21. ______
22. ______

15. $\frac{1}{4}\%$ of 80 =

16. $\frac{1}{5}\%$ of \$200 =

17. $\frac{3}{5}\%$ of \$152.78 =

18. $\frac{9}{10}\%$ of \$2,000 =

19. $\frac{4}{5}\%$ of 5,280 =

20. $\frac{1}{2}\%$ of \$424.80 =

21. $\frac{3}{10}\%$ of \$568.42 =

22. $\frac{2}{5}\%$ of 500 =

Name ______________________ Date ______________________

23. $\frac{3}{15}\%$ of \$900 =

24. $\frac{3}{5}\%$ of \$242.50 =

ANSWERS

23. ________

24. ________

25. ________

25. $\frac{3}{4}\%$ of 1,200 =

Solve these problems. Round to the nearest whole percent.

26. 25 is ______% of 50

27. 32 is ______% of 80

28. 75 is ______% of 200

29. 45 is ______% of 135

30. 42 is ______% of 90

31. 200 is ______% of 1,200

32. 320 is ______% of 950

33. 125 is ______% of 500

34. 500 is ______% of 2,500

35. 450 is ______% of 1,400

State the following as the nearest whole percent.

36. 15 out of 45 =

37. 20 out of 80 =

38. 60 out of 90 =

39. 45 out of 120 =

40. 75 out of 300 =

41. 200 out of 600 =

42. 50 out of 800 =

43. 8 out of 200 =

44. 60 out of 420 =

45. 50 out of 225 =

ANSWERS

36. ____________
37. ____________
38. ____________
39. ____________
40. ____________
41. ____________
42. ____________
43. ____________
44. ____________
45. ____________

Name ______________________ Date ______________________

Find the rate of increase (or decrease) in each of the following. Where applicable round percents to the nearest tenth percent.

	New Amount	Original Amount	Amount of Change	Percent Change
46.	$ 500	$ 300	$______	______%
47.	1,800	2,000	______	______%
48.	2,400	1,600	______	______%
49.	6,000	4,000	______	______%
50.	7,200	8,000	______	______%
51.	2,500	1,000	______	______%
52.	4,500	9,000	______	______%
53.	9,350	8,120	______	______%
54.	1,400	800	______	______%
55.	6,290	8,410	______	______%

Work Space

Name ______________________ Date ______________________

UNIT 41 Applications

ANSWERS

1. ____________
2. a. ____________
   b. ____________
3. ____________
4. ____________
5. ____________

1. The population of Baldwin County grew "one-half of one percent" during June 19X6. If the population was 2,456,800 at the beginning of June, what is the population at the end of June?

2. Last month, the economy's inflation rate was "nine-tenths of one percent." Write this as (a) a decimal and (b) as a percent.

3. Using decimal equivalents, prove that .1% and $\frac{1}{10}$% have the same value.

4. Which is larger, .8% or $\frac{7}{8}$%? How much larger? (Hint: convert the percents to decimals and compare the decimals.)

5. What is $\frac{3}{4}$% of $200?

6. Jim and Barbara Miller are looking for financing for a new home. They qualified for loans at two different banks. One bank quoted a rate of $10\frac{7}{8}\%$, and the other bank quoted a rate of 10.8%. Which rate is more attractive?

7. Find $\frac{9}{10}\%$ of $2,560.54.

8. James DeFoe earns $900 a month. But he only takes home $720. What percent of his earnings does he take home?

9. Of the 600 employees at Safeway Company, 320 are female. What percent of the workforce is female?

10. A suit that usually sells for $195 was marked down to $125. What is the percent reduction?

11. Sue Lang got a 94 on her first math test and an 85 on her second math test. What was her percent change?

ANSWERS

6. ______________
7. ______________
8. ______________
9. ______________
10. ______________
11. ______________

Name ______________________ Date ______________________

12. Last month, movie tickets went up from $4.50 a ticket to $5.00 a ticket. What is the percent increase?

ANSWERS

12. ________

13. ________

13. Buying a $60,000 house requires a $12,000 down payment. Give the down payment as a percent.

14. Fill out this financial statement for the Bellows Company. Round rates to the nearest whole percent.

Bellows Company
Position Statement
December 31, 19X2 and 19X3

	19X3	19X2	Change	% Change
ASSETS				
Cash	$ 12,000	$ 10,000	$ ______	______%
Receivables	28,000	26,000	______	______%
Inventory	30,000	24,000	______	______%
Supplies	1,000	1,200	______	______%
Equipment	90,000	72,000	______	______%
LIABILITIES				
Accounts Payable . .	$ 8,000	6,000	______	______%
Taxes Payable	1,200	1,400	______	______%
EQUITY				
Owner's Equity . . .	$151,800	125,800	______	______%

Work Space

UNIT 42 Aliquot Parts

When working with problems involving percents or decimals, calculations are often easier if you use a math shortcut called aliquot parts. An **aliquot part** is any number (whole or mixed) that divides evenly into another number without leaving a remainder. For example, \$.25 is an aliquot part of \$1.00 because \$.25 divides into \$1.00 exactly 4 times. In the same way, 20 is an aliquot part of 60; 5 is an aliquot part of 25; 9 is an aliquot part of 27; and $12\frac{1}{2}$ is an aliquot part of 100.

To illustrate how the use of aliquot parts can make problem solving easier, let's look at a couple of calculations.

EXAMPLE: What is the cost of 24 items if the cost per item is \$.25?

Regular Calculation	Use of Aliquot Part
24 × \$.25 120 48 \$6.00	$\frac{24}{1} \times \$\frac{1}{4} = \$6$ Since \$.25 divides into \$1.00 exactly 4 times, it is $\frac{1}{4}$ of a \$1.00.

As we can see from these calculations, multiplying by \$.25 is the same thing as multiplying by $\$\frac{1}{4}$ because \$.25 is one-fourth ($\frac{1}{4}$) of \$1.00. But it is much quicker to multiply by $\frac{1}{4}$.

Since use of an aliquot part will give you the same results as using its decimal equivalent, which should you use, the decimal or the aliquot part? The answer is simple—aliquot parts are best used when they make your calculations easier. Following is a table of the common aliquot parts of \$1.00; you should learn these.

COMMON ALIQUOT PARTS OF \$1.00

$\$\frac{1}{2}$	=	50 cents
$\$\frac{1}{3}$	=	$33\frac{1}{3}$ cents
$\$\frac{1}{4}$	=	25 cents
$\$\frac{1}{5}$	=	20 cents
$\$\frac{1}{6}$	=	$16\frac{2}{3}$ cents
$\$\frac{1}{7}$	=	$14\frac{2}{7}$ cents
$\$\frac{1}{8}$	=	$12\frac{1}{2}$ cents
$\$\frac{1}{10}$	=	10 cents
$\$\frac{1}{12}$	=	$8\frac{1}{3}$ cents
$\$\frac{1}{15}$	=	$6\frac{2}{3}$ cents
$\$\frac{1}{16}$	=	$6\frac{1}{4}$ cents
$\$\frac{1}{20}$	=	5 cents
$\$\frac{2}{3}$	=	$66\frac{2}{3}$ cents
$\$\frac{3}{4}$	=	75 cents
$\$\frac{2}{5}$	=	40 cents
$\$\frac{3}{5}$	=	60 cents
$\$\frac{4}{5}$	=	80 cents
$\$\frac{5}{6}$	=	$83\frac{1}{3}$ cents
$\$\frac{3}{8}$	=	$37\frac{1}{2}$ cents
$\$\frac{7}{8}$	=	$87\frac{1}{2}$ cents

Let's look at some examples to see how the use of aliquot parts can make problem solving easier.

EXAMPLE 1: What is the cost of 96 items if the cost per item is $12\frac{1}{2}$ cents?
Solution: $12\frac{1}{2}$ cents $= \$\frac{1}{8}$; so:

$$\frac{\overset{12}{\cancel{96}}}{1} \times \$\frac{1}{\underset{1}{\cancel{8}}} = \$12$$

EXAMPLE 2: Calculate the total price of 64 pens if one pen costs $37\frac{1}{2}$ cents.
Solution: $37\frac{1}{2}$ cents $= \$\frac{3}{8}$

$$\frac{\overset{8}{\cancel{64}}}{1} \times \$\frac{3}{\underset{1}{\cancel{8}}} = \$24$$

Aliquot parts can also be used for values that are greater than \$1.00. For example, what is the cost of 480 articles if one article costs $\$1.12\frac{1}{2}$?

$$\$1.12\tfrac{1}{2} = \$1\tfrac{1}{8} = \$\tfrac{9}{8} \quad \text{so} \quad \frac{\overset{60}{\cancel{480}}}{1} \times \$\frac{9}{\underset{1}{\cancel{8}}} = \$540$$

Alternate solution:

$$\begin{array}{r} 480 \times \$1 = \$480 \\ +\ 480 \times \$\frac{1}{8} = \quad 60 \\ \hline \$540 \end{array}$$

ALIQUOT PARTS OF 100%

While any number can be used as a base for aliquot parts, the two most common bases are \$1.00 and 100%. We saw in the above discussion how aliquot parts of \$1.00 can make calculations easier. Aliquot parts of 100% can also make many calculations easier. For example, what is $33\frac{1}{3}\%$ of 900? Working this out using a decimal is very hard because the decimal equivalent of $33\frac{1}{3}\%$ is a repeating decimal (.333333...). So, the following calculation using an aliquot part is much better.

$$33\tfrac{1}{3}\% = \tfrac{1}{3}, \quad \text{so} \quad \tfrac{900}{1} \times \tfrac{1}{3} = 300$$

CALCULATING ALIQUOT PARTS

If you do not recognize an aliquot part, and you do not have a table available, you can convert cents to aliquot parts of \$1.00 (or percents to aliquot parts of 100%) by following the technique shown in the following examples. In effect, you are simply changing a decimal, or a percent, to a fraction.

EXAMPLE 1: Convert $16\frac{2}{3}$ cents to an aliquot part of \$1.00.
Solution:

$$16\tfrac{2}{3}¢ = \frac{3 \times 16 + 2}{3 \times 100} = \frac{50}{300} = \frac{1}{6}$$

EXAMPLE 2: Convert $37\frac{1}{2}$ cents to an aliquot part of \$1.00.
Solution:

$$37\tfrac{1}{2}¢ = \frac{2 \times 37 + 1}{2 \times 100} = \frac{75}{200} = \frac{3}{8}$$

EXAMPLE 3: Convert $6\frac{1}{4}\%$ to an aliquot part of 100%.
Solution:

$$6\tfrac{1}{4}\% = \frac{4 \times 6 + 1}{4 \times 100} = \frac{25}{400} = \frac{1}{16}$$

Name ______________________________ Date ______________________________

UNIT 42 Exercises

Find the cost of each of these purchases. Use aliquot parts of $1.00 to make each calculation and write the aliquot part in the parentheses next to each unit price. Round off answers to even cents where necessary.

1. 160 units @ $.25 ( ) = $______

2. 300 units @ $\$.33\frac{1}{3}$ ( ) = $______

3. 252 units @ $\$.87\frac{1}{2}$ ( ) = $______

4. 900 units @ $\$.66\frac{2}{3}$ ( ) = $______

5. 540 units @ $.60 ( ) = $______

6. 144 units @ $\$.08\frac{1}{3}$ ( ) = $______

7. 600 units @ $\$.16\frac{2}{3}$ (   ) = \$______

8. 720 units @ $\$.87\frac{1}{2}$ (   ) = \$______

9. 648 units @ $\$.12\frac{1}{2}$ (   ) = \$______

10. 366 units @ $\$.83\frac{1}{3}$ (   ) = \$______

11. 800 units @ \$.80 (   ) = \$______

12. 336 units @ $\$.33\frac{1}{3}$ (   ) = \$______

13. 672 units @ \$.05 (   ) = \$______

Name ______________________ Date ______________________

14. 331 units @ $\$.14\frac{2}{7}$ ( ) = $______

15. 320 units @ $.40 ( ) = $______

Using aliquot parts of 100%, find the percentage (remember, B × R = P) in each of the following. Round answers to the nearest hundredth where necessary.

Base	Rate	Percentage
16. 800	$37\frac{1}{2}\%$	______
17. 961	50%	______
18. $600	$33\frac{1}{3}\%$	______
19. $352	$16\frac{2}{3}\%$	______
20. $720	$12\frac{1}{2}\%$	______
21. $480	$8\frac{1}{3}\%$	______
22. $301	$66\frac{2}{3}\%$	______
23. $916	$83\frac{1}{3}\%$	______
24. $200	5%	______
25. $618	75%	______

Using aliquot parts, calculate the extension (units × unit price) for each of the following. Round final answers to even cents where necessary.

Units	Unit Price	Extension
26. 1,200	$\$1.12\frac{1}{2}$	$______
27. 800	$\$1.16\frac{2}{3}$	______
28. 2,400	$\$2.08\frac{1}{3}$	______
29. 4,500	$\$3.06\frac{2}{3}$	______
30. 3,216	$\$1.06\frac{1}{4}$	______
31. 900	$\$2.66\frac{2}{3}$	______
32. 1,850	$\$41.37\frac{1}{2}$	______
33. 3,624	$\$2.16\frac{2}{3}$	______
34. 4,885	$1.20	______
35. 1,800	$1.80	______

Without referring to the chart in this unit, convert each of the following to aliquot parts of $1.00.

36. $.80 =

37. $\$.33\frac{1}{3}$ =

38. $.20 =

39. $\$.87\frac{1}{2}$ =

40. $\$.66\frac{2}{3}$ =

41. $\$.37\frac{1}{2}$ =

42. $\$.06\frac{2}{3}$ =

43. $\$.83\frac{1}{3}$ =

ANSWERS

36. ______________
37. ______________
38. ______________
39. ______________
40. ______________
41. ______________
42. ______________
43. ______________

Name ______________________ Date ______________________

44. $\$.98\frac{1}{3}$ =

45. $\$.16\frac{2}{3}$ =

Convert each of the following to aliquot parts of 100%.

46. $6\frac{1}{4}\%$ =

47. $87\frac{1}{2}\%$ =

48. 90% =

49. $33\frac{1}{3}\%$ =

50. 40% =

51. 75% =

52. $37\frac{1}{2}\%$ =

53. $12\frac{1}{2}\%$ =

54. 5% =

55. $3\frac{1}{3}\%$ =

ANSWERS

44. ____________

45. ____________

46. ____________

47. ____________

48. ____________

49. ____________

50. ____________

51. ____________

52. ____________

53. ____________

54. ____________

55. ____________

Work Space

Name ______________________ Date ______________________

UNIT 42 Applications

1. What is the total cost of 36 yards of fabric if one yard costs $\$1.37\frac{1}{2}$?

2. Todd McKing gets a commission of $6\frac{1}{4}\%$ on everything he sells. Last month his total sales were $56,800. What is his commission for the month?

3. If a certain cleaning solution costs $\$2.66\frac{2}{3}$ a pound, what would be the cost of 9 pounds?

4. Kelli Reese paid $\$.87\frac{1}{2}$ a quart for oil. How many quarts did she buy if her total bill came to $5.25?

5. John Lynn earns $\$4.12\frac{1}{2}$ an hour. How much does he earn in a week in which he worked 40 hours?

6. How much will 300 floor tiles cost at $\$.33\frac{1}{3}$ each?

ANSWERS

1. ____________
2. ____________
3. ____________
4. ____________
5. ____________
6. ____________

7. If heating oil sells for $\$1.06\frac{2}{3}$ a gallon, how much will 15 gallons cost?

ANSWERS

7. ____________

8. a. ____________

b. ____________

9. ____________

8. Of the 6,000 students at Maryvale College, $83\frac{1}{3}\%$ are from the United States. (a) How many students are from the United States? (b) How many students are from other countries?

9. Lakewood Community College has a total teaching faculty of 400 instructors. If $37\frac{1}{2}\%$ of the teaching faculty has a masters degree, how many instructors have masters degrees?

10. Fill in this invoice.

Ocala Supply Company
315 Central Avenue
Ocala, FL 32678

Sold to:

Florida Buildings Company
8000 College Parkway
Jacksonville, FL 32202

Item	Unit Cost	Extension
18 wrench sets	$\$3.66\frac{2}{3}$	$ ________
24 lock sets	$\$1.12\frac{1}{2}$	________
30 latch sets	$4.50	________
36 end holdings	$\$\ .16\frac{2}{3}$	________
56 place holders	$\$\ .87\frac{1}{2}$	________
Total		$ ________

UNIT 43

Trade Discounts

Producers and distributors of products often issue catalogs that show the **list price** or the **catalog price** of the merchandise they sell. A wholesale auto parts distributor, for example, could issue a catalog that describes the company's product line and gives a price for each product. These catalogs would then be distributed to the wholesaler's customers (mostly retail auto parts stores).

We can thus conclude that a seller's catalog serves two main purposes: (1) it gives a description of goods available, and (2) it gives the current prices of the goods. But what happens when the price of an item (or items) changes? Does this mean that the seller must pay to reprint the catalog so that the new prices will be shown in the catalog?

Usually it is too expensive to reprint a catalog each time the price of merchandise changes. Instead, many producers and distributors issue, in addition to catalogs, sheets of **trade discounts** that may be available. A trade discount is a percentage reduction in the list price of merchandise. When the price of merchandise changes, the seller can update the price by simply updating the discount sheet. For example, an item can be listed in a catalog for, say, $500 less a 10% discount. If the price of the item goes up, the 10% discount can be lowered, or dropped altogether.

Trade discounts also allow the seller greater flexibility when selling to different classes of customers, or when selling larger amounts. For example, sellers often offer a larger discount to customers who buy in large quantities.

The amount of a trade discount is determined as follows:

	List price of merchandise
×	Trade discount rate
=	Amount of trade discount

The amount of trade discount is then subtracted from the list price to get the **net price (list price − amount of trade discount = net price).** The net price is the amount the customer actually pays for the goods.

EXAMPLE 1: An item is listed at $800 less a 20% trade discount. Find the amount of the trade discount and the net price.

Solution:

Step 1: Calculate the amount of trade discount:

	$800	list price
×	.20	trade discount rate
	$160	amount of trade discount

Step 2. Calculate the net price:

	$800	list price
−	160	amount of trade discount
	$640	net price

EXAMPLE 2: What is the trade discount and net price of an item listed at $1,200 with a 40% trade discount?

Solution:

	\$1,200	list price
×	.40	trade discount rate
	\$480.00	amount of trade discount

	\$1,200	list price
−	480	amount of trade discount
	\$720	net price

When calculating the amount of trade discount, we are actually using the formula B × R = P:

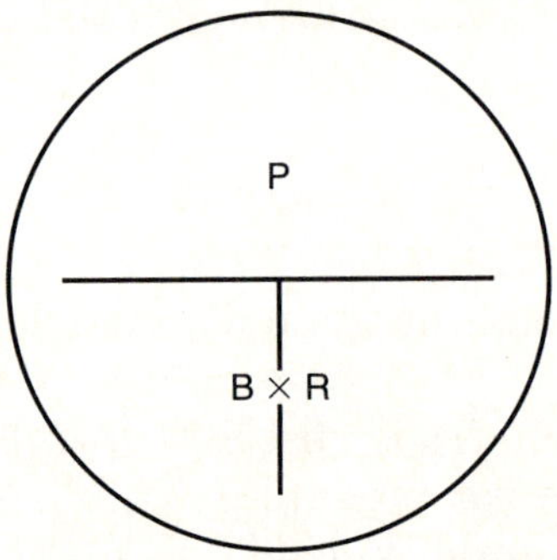

= B × R = P

B = List price of merchandise

R = Trade discount rate

P = Amount of trade discount

EXAMPLE: What is the trade discount on goods listed at \$600 less 40%?

Solution:

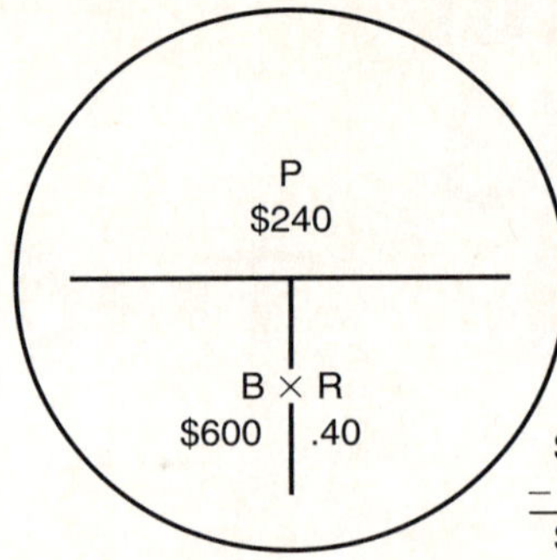

$B \times R = P$

$= \$600 \times .40 = \$240$

	\$600	list price
−	240	amount of trade discount
	\$360	net price

Helpful Hint: Another way to find the net price is to multiply the list price of merchandise by the **complement** of the trade discount rate. To get the complement of a rate, you subtract the rate from 100%. In the above example, for example, the complement of 40% is 60% (100% − 40%). You can then find the net price as follows:

List price	×	Complement of trade discount rate	=	Net price
\$600	×	.60	=	\$360

Remember: to work with a percent, you must change it to a decimal.

If you need to find the amount of trade discount, you can then subtract the net price from the list price:

	\$600	list price
−	360	net price
	\$240	amount of trade discount

Name ________________________________ Date ________________________________

UNIT 43 Exercises

Find the trade discount and net amount for each of the following.

List Price	Rate of trade discount	Amount of trade discount	Net price
1. $ 500	5%	$ _____	$ _____
2. $ 850	12%	$ _____	$ _____
3. $1,800	$10\frac{1}{2}$%	$ _____	$ _____
4. $2,540	25%	$ _____	$ _____
5. $9,000	$33\frac{1}{3}$%	$ _____	$ _____
6. $8,400	30%	$ _____	$ _____
7. $6,450	42%	$ _____	$ _____
8. $2,459	$16\frac{2}{3}$%	$ _____	$ _____
9. $4,200	28%	$ _____	$ _____
10. $9,560	31%	$ _____	$ _____
11. $6,280	$8\frac{1}{3}$%	$ _____	$ _____
12. $7,550	35%	$ _____	$ _____
13. $8,490	22%	$ _____	$ _____

	List Price	Rate of trade discount	Amount of trade discount	Net price
14.	$8,600	17%	$ ______	$ ______
15.	$6,240	27%	$ ______	$ ______
16.	$5,400	$37\frac{1}{2}\%$	$ ______	$ ______

Find the complement of the following rates.

17. 15%
18. 25%
19. 42%
20. 10%
21. 24%
22. 32%
23. 45%
24. 8%
25. 21%

ANSWERS

17. ____________
18. ____________
19. ____________
20. ____________
21. ____________
22. ____________
23. ____________
24. ____________
25. ____________

Name ______________________ Date ______________________

UNIT 43 Applications

ANSWERS

1. ________
2. a. ________
   b. ________
3. ________
4. ________

1. What is the net price of a personal computer listed at $1,200 with a trade discount of 20%?

2. Ledbetter Company bought a drill press that was listed at $48,500 with a trade discount of 25%. (a) What is the amount of trade discount? (b) What is the net price of the item?

3. Jim Swazey is thinking of buying a new air conditioning unit for his office. He found one with a list price of $2,800 and no discount. He found another with a list price of $3,400 and a trade discount of 12%. Which should he buy? Why?

4. Doyle Lowell bought a computer keyboard and got a $50 trade discount. If the keyboard had a list price of $300, what rate of discount did Doyle get? (Hint: B = $300, P = $50, find R.)

5. What is the net price of a grandfather clock that has a list price of $1,800 and a trade discount of $33\frac{1}{3}$%?

ANSWERS

5. __________

6. __________

7. __________

8. a. __________

   b. __________

6. Carlo Rojas paid $1,200 for an item that was listed for $1,800. What rate of discount did he get? (Hint: $1,200 = the complement of the trade discount.)

7. Campus Bookstore paid $7,500 net for textbooks this semester. If a 25% trade discount was given by the publisher, what was the original list price?

8. Jones Tire Company got a $1,500 discount on its last purchase. If the rate of discount was 15%, what was (a) the list price, (b) the net price?

UNIT

44

Two or More Trade Discounts in a Series

In Unit 43, you learned that a trade discount is a percentage reduction in the list price of merchandise. You also learned that a trade discount is calculated as follows:

List price × trade discount rate
= amount of trade discount

You then subtract the amount of trade discount from the list price to get the net price.

To allow for greater flexibility in pricing items, sellers sometimes offer two or more discounts in a series, for example, 20% and 10%. Such a discount is called a **chain discount** or a **series discount.** Chain discounts are especially helpful when the seller sets the discounts available for different classes of customers, different types of products, special promotional sales, seasonal fluctuations, and the like.

The rates in a chain discount are calculated on a successive basis, that is, the first discount is calculated and subtracted; the second discount is taken on what is left. And if there is a third discount, it is taken on the amount left after the second discount is calculated and subtracted, and so on.

EXAMPLE: Calculate the net price of an item listed at $800 with a chain discount of 20% and 10%.

Solution:

Step 1	Step 2	Step 3
$800	$800	$640
× .20	− 160	− 64
$160	$640	$576 net price
	× .10	
	$ 64	

Note that the second discount (Step 2) was calculated on the remainder after the first discount was subtracted. The second discount ($64) was then subtracted to yield a net price of $576. It does not matter what order you calculate the discounts because they are figured on a successive basis. Thus, you could have calculated them as 10% and 20% (rather than as 20% and 10%) and you would get the same net price. However, **you can never add the rates together,** because each discount is calculated on a successively smaller base. Let's look at another example.

EXAMPLE: Calculate the net price of an item listed at $1,500 with a chain discount of 20%, 10%, and 10%.

Step 1	Step 2
$1,500	$1,500
× .20	− 300
$300.00	$1,200
	× .10
	$ 120

Step 3	Step 4
$1,200	$1,080
− 120	− 108
$1,080	$ 972 net price
× .10	
$ 108	

We can now add the three discounts together to get the total discount: $300 + $120 + $108 = $528.

EQUIVALENT SINGLE DISCOUNT

A series of discounts can be changed to a single discount that has the same value as all discounts in the series. That is, you can convert two or more discount rates in a series to an equivalent single discount rate. To illustrate, let's look again at the preceding example in which we had an item with a list price of $1,500 with a chain discount of 20%, 10%, and 10%.

To find the equivalent single discount rate:

Step 1: Subtract each discount from 100%; then find the product of these differences (remember, you must change percents to decimals to work with them):

$$\begin{array}{r}1.00\\-\ \underline{.20}\\.80\end{array}\times\begin{array}{r}1.00\\-\ \underline{.10}\\.90\end{array}\times\begin{array}{r}1.00\\-\ \underline{.10}\\.90\end{array}=.648$$

do not round off

Step 2: Take the result from Step 1 (.648 in this case) and subtract it from 100%.

We added a zero so that we could make subtraction.

$$\begin{array}{r}1.000\\-\ \underline{.648}\\.352\end{array}$$ **equivalent single discount**

Now that we found our equivalent single discount, we can multiply it times the list price to get the amount of discount:

$1,500	list price
× .352	equivalent single discount
$ 528	amount of discount

Now, subtract the amount of discount, $528, from the list price to get the net price:

$1,500	
− 528	
$ 972	net price

Note that using the equivalent single discount yielded the same net price, $972, as calculating the discounts successively, but with less work. The equivalent single discount is especially helpful when companies often buy items that have the same chain discount.

NET PRICE EQUIVALENT RATE

As you just learned, you can multiply the equivalent single discount rate times the list price of an item to get the trade discount. You then subtract the trade discount from the list price to get the net price.

We can alter this process just a bit and get the same results. Let's go back to our earlier example of a $1,500 item offered with a chain discount of 20%, 10%, and 10%. Remember, the first step in calculating an equivalent single discount rate is to subtract each discount from 100% (1.00) and multiply these differences together, which we did:

$$\begin{array}{r}1.00\\\underline{.20}\\.80\end{array}\times\begin{array}{r}1.00\\\underline{.10}\\.90\end{array}\times\begin{array}{r}1.00\\\underline{.10}\\.90\end{array}=.648$$

Now, stop with this rate (instead of subtracting it from 100% as we did earlier). This rate is equivalent to the net price. So, multiplying it times the list price will yield the net price:

List price	×	net price equivalent rate	=	net price
$1,500	×	.648	=	$972

To get the amount of trade discount, we can subtract the net price from the list price:

	$1,500	list price
−	972	net price
	$ 528	amount of trade discount

Let's look at another example.

EXAMPLE: Using the net price equivalent rate, find the net price and amount of trade discount of an item listed at $1,200 with a chain discount of 20% and 5%.

Solution:

First, find the net price equivalent rate:

1.00	1.00	
.20	.05	
.80 ×	.95 =	**.76 net price equivalent rate**

Next, multiply the net price equivalent rate times the list price to get the net price:

	$1,200	list price
×	.76	net price equivalent rate
	$ 912	net price

Finally, subtract the net price from the list price to find the amount of trade discount:

	$1,200	
−	912	
	$ 288	amount of trade discount

Answers:

Net price = $912
Trade discount = $288

Remember.

1. A chain discount is two or more discounts in a series.
2. You never add discounts in a series together because the discounts are taken on successively smaller bases.
3. Multiplying the list price by the equivalent single discount rate yields the trade discount.
4. Multiplying the list price by the net price equivalent rate yields the net price.

Work Space

Name ______________________ Date ______________________

UNIT 44 Exercises

Find the equivalent single discount rate for each of the following. Round off decimals to four places, where necessary.

ANSWERS

1. ____________
2. ____________
3. ____________
4. ____________
5. ____________
6. ____________

1. 30%, 20%, and 10%

2. 20% and 10%

3. 20% and 20%

4. 20%, 25%, 5%, and 10%

5. 20%, 20%, and 10%

6. 10%, 10%, and 10%

7. 5%, 10%, and 5%

ANSWERS

7. ______
8. ______
9. ______
10. ______
11. ______
12. ______

8. $33\frac{1}{3}$%, 15%, and 5%

9. $7\frac{1}{2}$%, 10%, and 10%

10. 40%, 20%, and 5%

11. 10% and 10%

12. 5% and $2\frac{1}{2}$%

Name ______________________ Date ______________________

Using equivalent single discount rates, find the amount of trade discount and the net price for each of the following. Round to four places, where necessary, and round all dollar amounts to even cents.

List price	Chain discount	Amount of trade discount	Net price
13. $ 800	20% and 12%	$_____	$_____
14. $ 900	$33\frac{1}{3}$% and 15%	_____	_____
15. $ 1,450	25%, 15%, and 5%	_____	_____
16. $ 2,500	40% and 10%	_____	_____
17. $ 6,400	10%, 5%, and 2%	_____	_____
18. $ 9,800	20%, 20%, and 10%	_____	_____
19. $12,500	25%, 20% and 5%	_____	_____
20. $60,450	15%, 12%, and 5%	_____	_____

Work Space

Name ______________________ Date ______________________

UNIT 44 Applications

In the following problems, round all rates, where necessary, to four places and round all dollar amounts to even cents.

1. Find the net price of a walk-in cooler if the cooler has a list price of $9,000 and is offered with a chain discount of $33\frac{1}{3}$%, 20%, and 10%.

2. Tom Jarrel is thinking about buying a freezer. He found two units that he liked. One cost $1,200 with a chain discount of 20%, 10%, and 5%. The other cost $1,255 with a chain discount of 20%, 12%, and 4%. Which should he buy? Why?

3. Edge Manufacturing Company sold a set of tools to American Buildings Company. The list price of the tools was $1,400 with a chain discount of 10%, 10%, and 2%. What was (a) the amount of trade discount, and (b) the net price?

ANSWERS

1. ______

2. ______

3. a. ______

   b. ______

4. Lee King bought a welder and got a $171 trade discount. If the welder was offered with a chain discount of 10% and 10%, what was the list price? (Hint: $171 = P; 10% and 10% = R; find B)

ANSWERS

4. ____________

5. ____________

6. ____________

5. An item with a list price of $1,500 was bought for $1,200. What rate of trade discount was received?

6. What is the amount of trade discount and the net price of an item listed at $12,000 with a chain discount of $33\frac{1}{3}$%, 20%, and 10%? (Helpful hint: use aliquot parts instead of decimals.)

Name ______________________ Date ______________________

7. Which of the following chain discounts will yield the greater discount: (a) 20%, 20%, and 10% or (b) 20%, 18%, and 12%? How much more? (Hint: convert the discounts to equivalent single discount rates and compare the single rates.)

ANSWERS

7. a. ________

b. ________

8. Differentiate between the *equivalent single discount rate* and the *net price equivalent rate.*

Work Space

UNIT

45

Cash Discounts

In Units 43 and 44, we discussed trade discounts, which, as you recall, are deductions from a catalog or list price. Another type of discount often given by sellers is the **cash discount,** which is a discount offered to encourage customers to make prompt and early payment. The rate of a cash discount and the time the customer has to take advantage of the cash discount are part of the **credit terms** and will appear on the invoice.

A common form of cash discount is **2/10,n/30** (read as **two ten, net thirty**). This means that a 2% discount can be taken if the invoice is paid within 10 days from the invoice date; if payment is not made within 10 days, the full amount must be paid within 30 days from the invoice date. If payment is not made within 30 days, the debt is past due and the seller can charge interest or a late fee. Let's look at an example of an invoice that is paid within the discount period.

EXAMPLE: An item with an invoice price of $800 was purchased on June 2 with terms of 2/10,n/30. If payment was made on June 12, what was the net amount paid?

Solution:

Step 1: Calculate the cash discount:

	$800	invoice price
×	.02	
	$ 16	cash discount

Step 2: Subtract the cash discount from the invoice price.

	$800	
−	16	
	$784	net amount paid

Cash discounts are only taken on the price of merchandise; they *are not* taken on charges for such things as freight and packing. If charges for freight and packing are included in the invoice, they must be deducted before the cash discount is taken. They are then added back in order to get the total amount due.

EXAMPLE: Phillips Company received goods that were invoiced for $780. Included in the invoice price was a freight charge of $30. What is the amount due if the invoice carried terms of 2/10,n/30 and payment was made within the discount period?

Solution:

Step 1: Subtract the freight charge from the invoice price:

	$780	total of invoice
−	30	freight charge
	$750	amount subject to cash discount

Step 2: Multiply the rate of cash discount times the amount subject to the discount:

	$750	
×	.02	rate of cash discount
	$ 15	amount of cash discount

Step 3: Subtract the cash discount and add back the freight:

	$750	
−	15	
	$735	
+	30	freight charge
	$765	total amount due

Terms	Read as	Meaning
n/30	"net thirty"	No discount. Payment is due within 30 days from the invoice date.
c.o.d.	"cash on delivery"	Payment is due when the merchandise is delivered to the buyer.
2/10,n/e.o.m.	"two ten, net end of month"	A 2% discount can be taken if payment is made by the 10th day of the month that **follows** the month of sale. If the discount is not taken, payment must be made within 20 days after the discount period.
2/10,n/30 r.o.g.	"two ten, net thirty receipt of goods"	A 2% discount can be taken if payment is made within 10 days **after** receipt of the goods; full amount is due between day 11 and day 30 if payment is not made within discount period.

OTHER COMMON TERMS OF SALE

Terms of 2/10,n/30 are very common. But there are other common terms of sale. Look at the above chart to see some of them.

Now, before starting the exercises, let's look at a couple of other examples.

EXAMPLE 1: On May 18, Rodriguez Company bought goods with an invoice price of $900. The invoice carried terms of 2/10, n/e.o.m. and payment was made on June 9. What is the net amount due?

Solution:
The discount applies because payment was made within 10 days after the end of the month:

$$\$900 \times .02 = \$18; \$900 - \$18 = \$882$$

EXAMPLE 2: Taylor Supply Company bought $6,000 worth of merchandise on July 6. The merchandise was delivered on July 11 and payment was made on July 21. What is the net amount paid if terms of sale were 2/10, n/30 r.o.g.?

Solution:
The discount applies because payment was made on July 21, which is 10 days after the merchandise was received:

$$\$6{,}000 \times .02 = \$120; \$6{,}000 - \$120 = \mathbf{\$5{,}880}$$

Remember.

1. A cash discount is offered to encourage prompt payment of an invoice; it is not automatic.
2. The cash discount period starts running with the date of the invoice, not the actual date of delivery.
3. Only the actual price of goods is subject to a cash discount. And even though they must be added back after the cash discount is taken, additional charges for such things as freight and packing are not subject to the discount.

Name ______________________________ Date ______________________________

UNIT 45 Exercises

Find the amount of cash discount and the net amount paid for each of the following. Round all dollar amounts to even cents.

	Invoice Amount	Invoice Date	Terms	Date of Payment	Cash Discount	Net Amount Paid
1.	$ 1,800	Jan. 4	2/10,n/30	Jan. 14	$_____	$_____
2.	$ 4,500	Mar. 12	1/10,n/30	Mar. 18	$_____	$_____
3.	$ 600	Oct. 28	3/20,n/90	Nov. 15	$_____	$_____
4.	$ 1,250	June 30	3/10,2/20,n/45	July 20	$_____	$_____
5.	$12,500	Aug. 12	4/10,n/30	Aug. 22	$_____	$_____
6.	$62,578	Dec. 1	2/10,1/30,n/90	Dec. 31	$_____	$_____
7.	$22,400	Feb. 3	2/10,1/20,n/30	Mar. 5	$_____	$_____
8.	$34,800	Mar. 7	2/10,n/e.o.m.	Apr. 9	$_____	$_____
9.	$12,800	June 14	1/10,n/90	June 24	$_____	$_____
10.	$45,675	Oct. 6	2/10,n/e.o.m.	Nov. 10	$_____	_____

Work Space

Name ______________________ Date ______________________

UNIT 45 Applications

1. Goods with an invoice price of $525 were bought on June 18. The invoice carried terms of 2/10,n/30 and payment was made on June 28. What is (a) the amount of cash discount and (b) the net amount paid?

ANSWERS

1. a. __________

   b. __________

2. a. __________

   b. __________

3. __________

2. On March 2, Carlsbad Company bought $3,000 worth of merchandise from Lincoln Company. Terms of sale were 3/10,2/20, n/45. What was the net amount paid if Carlsbad made payment on (a) March 22, (b) March 11?

3. On July 14, Simmers Company received goods invoiced at $2,740 from Tyler Company. The invoice was dated July 12 and included a freight charge of $35. If terms of payment were 2/10,n/30, and payment was made on July 23, what was the net amount paid?

4. Mills, Inc. received a shipment on October 15. The invoice, dated October 14, carried a list price of $1,800 subject to a 10% trade discount and terms of 2/10,n/30. Also shown on the invoice was a freight charge of $25 and a packing charge of $10. If Mills made payment on October 24, what was the net amount paid?

ANSWERS

4. ______

5. Find the net amount to be paid for each of the following invoices:

	Invoice Date	List Price	Trade Discount	Terms	Date of Payment	Net Amount Paid
A.	Jan 15	$4,000	none	2/10,n/30	Jan. 25	$______
B.	Apr. 10	$5,600	10% and 10%	3/10,2/15,n/30	Apr. 24	$______
C.	May 28	$7,200	20%, 10%, and 5%	2/15,n/90	July 22	$______
D.	Feb. 4	$9,000	$33\frac{1}{3}$%	3/20,n/45	Feb. 24	$______

Name ______________________ Date ______________________

6. Goods with an invoice price of $7,400 were received on November 12. The date of the invoice was November 10 and terms of sale were 3/10,2/20,1/30,n/90. What was the net amount paid if payment was made on:

   a. November 22

   b. November 20

   c. November 30

   d. December 1

   e. December 31

   f. January 15 (of the following year)

ANSWERS

6. a. ________
   b. ________
   c. ________
   d. ________
   e. ________
   f. ________

ANSWERS

7. a. ________

b. ________

8. ________

7. A drill press with an invoice price of $34,800 was received on June 5. The date of the invoice was June 2 and terms of sale were 2/10,n/30 r.o.g. What was the net amount paid if payment was made on (a) June 15, (b) June 22?

8. Goods invoiced at $800 were received on July 14. Terms of sale were 2/10,n/e.o.m. What was the net amount paid if payment for the goods was made on August 10?

UNIT 46 Computing Markup and Selling Price

To start our discussion of markup and selling price, let's take an example.

> Mark Price is thinking about driving to a nearby state to buy a used car. Mark plans to fix up the car and resell it for a profit. Mark can buy the car for $800 and he calculates that he will have to spend another $200 to get the car in condition to sell. Mark will also spend $10 for gas to get to the car and have someone drive it back. Of course, Mark wants to make as much profit as he can on the transaction. But he knows that he must set a price that is realistic. So, he first calculates his total investment in the car:
>
> | | |
> |---|---|
> | Cost . . . . | $ 800 |
> | Fixing up expenses . . . | 200 |
> | Gas . . . . | 10 |
> | Total . . . . | $1,010 |
>
> Mark now realizes that to make a profit, he must sell the car for more than his $1,010 investment in the car. Mark checks with several used car dealers and several banks to get the fair market value of a car like this. He finds that cars of this make and age are selling for around $1,600. Thus, Mark decided to set a selling price of $1,595.

In setting his selling price, Mark considered two factors: (1) how much he paid for the car (cost) and (2) how much he must add to the cost to cover expenses and make a profit. This second factor is called **markup;** it is the amount added to the cost of goods to get the selling price.

Mark calculated the selling price of his car by adding a definite dollar amount to the cost of the car. His markup was the difference between his cost and the selling price. In business, markup is usually calculated by using a rate. There are two common bases used to determine markup: **markup based on cost** and **markup based on the selling price.** We will look at both methods in the discussion that follows.

MARKUP BASED ON COST

To calculate the amount of markup when cost is the base, you multiply the cost of an item times the markup rate. Then, you add the markup to the cost to get the selling price. In formula form, this looks like this:

Cost of an item	Cost of an item
× Markup rate	+ Markup
= Markup	= Selling price

EXAMPLE: On Monday morning, Sterling Clothing Store received a shipment of men's suits. Each suit had a cost of $90 and Sterling marks up all merchandise 40% of cost. What is the selling price of the suits?

Solution:

Step 1: Calculate the markup:

	$90	cost
×	.40	markup rate
	$36	markup

Step 2: Add the markup, $36, to the cost to get the selling price:

	$ 90	cost
+	36	markup
	$126	**selling price**

Each business must decide what rate to use when marking up goods. Some merchandise, such as groceries, has a low markup because of high sales volume and heavy competition. Other merchandise, such as jewelry, usually has a much higher markup because the sales volume is usually low. Let's look at another example of markup based on cost.

EXAMPLE: What is the selling price of an item with a cost of $600 and a markup rate of $33\frac{1}{3}\%$ based on cost?
Solution:

Cost	×	Markup rate	=	Markup
$600	×	$\frac{1}{3}$	=	$200

Remember: the aliquot part of $33\frac{1}{3}\%$ is $\frac{1}{3}$.

Cost	+	Markup	=	Selling price
$600	+	$200	=	**$800**

FINDING THE RATE OF MARKUP BASED ON COST

In some problems, you will be given the selling price and the cost price, and you will be asked to find the rate of markup. To find the rate of markup when cost is the base, you divide the amount of markup by the cost price. We can express this in formula form:

$$\frac{\text{Amount of markup}}{\text{Cost}} = \text{Rate of markup based on cost}$$

Remember: a horizontal line over two numbers means to divide.

EXAMPLE: Goods costing $500 were marked to sell for $600. What is the rate of markup based on cost?

Solution:

Step 1: Determine the amount of markup by subtracting the cost from the selling price:

$600	Selling price
− 500	Cost
$100	Amount of markup

Step 2: Divide the amount of markup by the cost:

$$\frac{\$100}{\$500} = .2 \text{ or } 20\%$$

Proof:		
	$500	cost
	× .20	markup rate
	$100	amount of markup

MARKUP BASED ON THE SELLING PRICE

Markup can also be based on the selling price, or the retail price, of goods. Under this assumption, the selling price is calculated using this formula:

$$\frac{\text{Cost}}{100\% - \text{the markup rate}} = \text{Selling price of goods}$$

Then, to get the amount of markup, you subtract the cost of the goods from the selling price.

EXAMPLE 1: An item with a cost of $600 is marked up 40% of the selling price. Determine the selling price.
Solution:

$$\frac{\text{Cost}}{100\% - \text{the markup rate}} = \frac{\$600}{1.00 - .40} = \frac{\$600}{.60} = \mathbf{\$1,000}$$

Remember: change your percents to decimals.

Proof: $1,000 × .40 = $400 markup
$1,000 − $400 = $600 cost

EXAMPLE 2: An item with a cost of $420 is marked up 30% of the selling price. Determine the selling price and the amount of markup.
Solution:
Determine the selling price:

$$\frac{\$420}{1.00 - .30} = \frac{\$420}{.70} = \mathbf{\$600}$$

Now, subtract the cost from the selling price to get the markup:

$600	Selling price
− 420	Cost
$180	**markup**

Proof: $600 × .30 = $180 markup
$600 − $180 = $420 cost

FINDING THE RATE OF MARKUP BASED ON SELLING PRICE

To find the rate of markup based on selling price, divide the amount of markup by the selling price. We can state this in formula form, as follows:

$$\frac{\text{Amount of markup}}{\text{Selling price}} = \text{Rate of markup based on selling price}$$

EXAMPLE: Twin City Furniture Company sold a bed for $300 that cost $225. What was the rate of markup based on the selling price?
Solution:

$300 − $225 = $75 markup

$75 ÷ $300 = 25% rate of markup based on selling price

Proof: $300 × .25 = $75 markup

Work Space

Name ______________________ Date ______________________

UNIT 46 Exercises

Find the markup based on cost and the selling price for each of the following.

Cost	Markup Rate on Cost	Amount of Markup	Selling Price
1. $ 400	25%	$______	$______
2. 575	30%	______	______
3. 820	10%	______	______
4. 1,240	27%	______	______
5. 1,500	$33\frac{1}{3}$%	______	______
6. 3,000	$37\frac{1}{2}$%	______	______
7. 6,200	40%	______	______
8. 7,852	35%	______	______
9. 1,455	18%	______	______
10. 9,800	$12\frac{1}{2}$%	______	______

Find the selling price and the markup based on selling price for each of the following. Round dollar amount to even cents, where necessary.

	Cost	Markup Rate of Selling Price	Selling Price	Amount of Markup
11.	$ 800	20%	$_____	$_____
12.	1,200	40%	_____	_____
13.	784	30%	_____	_____
14.	228	25%	_____	_____
15.	600	$33\frac{1}{3}\%$	_____	_____
16.	2,480	15%	_____	_____
17.	4,400	22%	_____	_____
18.	3,200	60%	_____	_____
19.	2,700	10%	_____	_____
20.	6,400	36%	_____	_____

Name ____________________ Date ____________________

For each of the following, find the amount of markup. Then find the rate of markup based on cost and the rate of markup based on selling price. Round rates to the nearest tenth percent, where necessary.

	Selling Price	Cost	Amount of Markup	Rate of Markup Based on Cost	Rate of Markup Based on S.P.
21.	$ 800	$ 500	$_____	_____%	_____%
22.	345	210	_____	_____	_____
23.	872	650	_____	_____	_____
24.	1,450	900	_____	_____	_____
25.	3,000	2,200	_____	_____	_____
26.	4,200	3,500	_____	_____	_____
27.	8,000	6,000	_____	_____	_____
28.	4,500	3,000	_____	_____	_____

Work Space

Name ______________________ Date ______________________

UNIT 46 Applications

In the following application problems, round all rates to three places and all dollar amounts to even cents, where necessary.

1. Goods costing $672 are marked up 25% of cost. What is the selling price?

2. Ben's Appliance Company bought a refrigerator for $890 and marked it up 30% of cost. What is (a) the amount of markup, and (b) the selling price?

3. A store bought a couch for $600 and marked it up 40% of the selling price. What is (a) the selling price, and (b) the markup?

4. Goods costing $6,500 were marked to sell for $8,000. (a) What was the rate of markup based on cost? (b) What was the rate of markup based on selling price?

ANSWERS

1. ________
2. a. ________
   b. ________
3. a. ________
   b. ________
4. a. ________
   b. ________

5. The cost of merchandise in a small department store was $24,800. The selling price totaled $32,600.
   a. What was the markup?

   b. What was the rate of markup based on cost?

   c. What was the rate of markup based on selling price?

ANSWERS

5. a. ________
   b. ________
   c. ________
6. a. ________
   b. ________
7. a. ________
   b. ________

6. A house was constructed at a cost of $60,000. It was then sold for $90,000. (a) What is the rate of gain based on cost? (b) What is the rate of gain based on selling price?

7. A manufacturer sold goods to a wholesaler (a middleman) for $40,000. The wholesaler marked the goods up 20% of cost and sold them to a retailer. The retailer then marked the goods up 30% of the selling price and placed them on the shelf. (a) What was the cost of the goods to the retailer? (b) At what price did the retailer mark the goods for resell?

Name ______________________________ Date ______________________________

8. A sporting goods store bought baseball gloves for $144 a dozen. What is the price of each glove if they were marked up 40% of the selling price?

ANSWERS

8. ____________

Work Space

UNIT 47 Computing Markdown

Businesses often mark down the price of merchandise because of such things as promotional sales, seasonal changes, style changes, and inventory clearance. This reduction in the original selling price of merchandise is called **markdown,** and the selling price after the markdown is called the **new selling price:**

Original selling price − Markdown
= new selling price

Markdown can be expressed as a definite dollar amount. For example, a \$500 item can be marked down by \$50, which then gives a new selling price of \$450 (\$500 − \$50). Or, markdown can be expressed as a rate (percent).

EXAMPLE: An item with a selling price of \$800 is marked down by 20%. What is the new selling price?
Solution:

Original selling price . .	\$800	\$800
Markdown rate	× .20	− 160
Amount of markdown . .	\$160	**\$640**

new selling price

If a problem involves both a markup and a markdown, *the markup must be computed first.*

EXAMPLE: What is the new selling price of a sofa costing \$300, marked up 25% of cost, and then marked down by 40%?

Solution:

\$300	Cost
+ 75	Markup (\$300 × .25)
\$375	Original selling price
− 150	Markdown (\$375 × .40)
\$225	New selling price

Prices sometimes tend to move up and down. Any increase in price after a markdown is called a **markdown cancellation.** We can express this in formula form as follows:

Markdown − Markdown cancellation
= Net markdown

Now, we find the final selling price as follows:

Original selling price − Net markdown
= Final selling price

EXAMPLE: An item with an original selling price of \$80 was marked down by 30%, and later raised by 10%. What is the final selling price and the net markdown?
Solution:

\$80.00	Original selling price
− 24.00	Markdown (\$80 × .30)
\$56.00	
+ 5.60	Markdown cancellation (\$56 × 10%)
\$61.60	**Final selling price**

Net markdown = \$24.00 − 5.60 = \$18.40

Note: you can also find the final selling price ($61.60) by subtracting the net markdown ($18.40) from the original selling price.

FINDING THE RATE OF MARKDOWN

Sometimes you will be given the original selling price and the new selling price, and you will be asked to calculate the rate (percent) of markdown. To do this, you divide the amount of markdown by the original selling price. In formula form, this looks like this:

$$\frac{\text{Markdown}}{\text{Original selling price}} = \text{Rate of markdown}$$

EXAMPLE: Goods with an original selling price of $6,000 were marked to sell for $4,500. What was the rate of markdown?

Solution:

Step 1: Find the amount of markdown:

$6,000	Original selling price
− 4,500	New selling price
$1,500	Markdown

Step 2: Divide the markdown by the original selling price:

$$\frac{\$1{,}500}{\$6{,}000} = .25 = 25\%$$

Proof: $6,000 × .25 = $1,500

Name ______________________ Date ______________________

UNIT 47 Exercises

For each of the following, compute the markdown and the new selling price. Round all dollar amounts to even cents, where necessary.

	Original Selling price	Rate of Markdown	Markdown	New Selling Price
1.	$ 800	25%	$_____	$_____
2.	950	30%	$_____	$_____
3.	780	10%	$_____	$_____
4.	1,200	$33\frac{1}{3}$%	$_____	$_____
5.	4,000	15%	$_____	$_____
6.	6,450	40%	$_____	$_____
7.	8,455	10%	$_____	$_____
8.	8,000	$37\frac{1}{2}$%	$_____	$_____
9.	12,800	32%	$_____	$_____
10.	15,600	45%	$_____	$_____
11.	45,200	28%	$_____	$_____
12.	14,250	36%	$_____	$_____

Find the rate of markdown for each of the following. Round rates to the nearest tenth percent, where necessary.

	Original Selling Price	New Selling Price	Rate of markdown
13.	$ 600	$ 450	____%
14.	900	600	____%
15.	1,400	1,200	____%
16.	2,500	2,000	____%
17.	4,250	2,750	____%
18.	6,240	4,840	____%
19.	5,000	4,000	____%
20.	3,800	3,200	____%

Name ______________________ Date ______________________

UNIT 47 Applications

ANSWERS

1. a. ________
   b. ________
   c. ________
2. ________
3. a. ________
   b. ________
   c. ________
   d. ________

1. An item with a cost of $600 was marked up 40% of the selling price. It was then marked down 20% and sold. What was (a) the original selling price, (b) the markdown, and (c) the new selling price?

2. Goods with an original selling price of $4,200 were marked down 25%. What is the new selling price?

3. An item with a cost of $500 was marked up 40% of cost. It was then marked down 15% and sold. Find:

   a. markup

   b. the original selling price

   c. markdown

   d. new selling price

4. Goods with an original selling price of $2,000 were marked to sell for $1,800. What was the rate of markdown?

ANSWERS

4. ____________
5. ____________
6. ____________
7. a. ____________
   b. ____________
8. ____________

5. If a 20% markdown amounts to $27, what was the original selling price? (Hint: P = $27, R = 20%, find B)

6. The new selling price of an item is $48. What was the rate of markdown if the original selling price was $60?

7. A bicycle is marked down $40, which is 40% of the original selling price. (a) What was the original selling price? (b) What is the new selling price?

8. Harry's Clothing Store marked shoes down from $60 to $40. What was the rate of markdown?

UNIT 48

Sales Taxes

Like private businesses, the various forms of governments (federal, state, and local) must have money (revenue) in order to operate. One source of operating revenue for state governments is the **sales tax,** which is a tax on the retail price of goods (and some services).

It is the seller's responsibility to collect the sales tax from the buyer and pay it to the government. Sales tax rates vary from state to state, and from county to county, because it is also common for local governments to levy (charge) a sales tax. Goods subject to the tax also vary from state to state. For example, in some states there is no sales tax on food items. For our problems, we will assume a sales tax rate of 7% and that all items sold at retail are subject to the tax.

To calculate sales tax on whole dollar amounts, you multiply the sales tax rate times the amount of sale.

EXAMPLE: Find the sales tax on a $500 item in an area with a 7% sales tax rate.
Solution:

	$500	Amount of sale
×	.07	Sales tax rate
	$ 35	**Sales tax**

To calculate sales tax on sales less than $1.00, use the following table:

7% SALES TAX TABLE

Amount of Sale	Tax
$0.00–.09	$.00
.10–.20	.01
.21–.33	.02
.34–.47	.03
.48–.62	.04
.63–.76	.05
.77–.91	.06
.92–.99	.07

EXAMPLE 1: What is the sales tax on a sale of $.69?
Solution:

The amount of the sale, $.69, falls in the bracket $.63–$.76. Thus, the tax is $.05.

EXAMPLE 2: What is the sales tax and the total amount due on a sale of $.95?

Tax = $.07; $.95 + .07 = $1.02 amount due

To calculate sales tax on sales that involve dollar and cent amounts, figure 7% on the whole dollar amounts and use the table for sales under $1.00.

EXAMPLE 1: Find the sales tax on a sale of $10.50.

Solution:

Step 1: Find the tax on the whole dollar amount:

$$\$10 \times .07 = \$.70$$

Step 2: Use the table to find the tax on $.50:

$$\$.48 - .62 = \$.04$$

Step 3: Add the two amounts: $.70 + $.04 = **$.74**

EXAMPLE 2: Find the sales tax on a sale of $125.81.

Solution:

$125 × .07	=	$8.75
Tax on $.81 (table)	=	.06
Total tax		$8.81

Name ______________________ Date ______________________

UNIT 48 Exercises

Find the sales tax and the total amount due for each of the following sales. All sales are subject to a 7% sales tax. Use the table presented in the unit where needed.

	Amount of Sale	Amount of Sales Tax	Total Amount Due
1.	$ .69	$______	$______
2.	.99	______	______
3.	.25	______	______
4.	1.29	______	______
5.	12.95	______	______
6.	139.95	______	______
7.	525.00	______	______
8.	900.00	______	______
9.	234.58	______	______
10.	107.99	______	______
11.	4,000.00	______	______
12.	3,569.69	______	______

	Amount of Sale	Amount of Sales Tax	Total Amount Due
13.	6,255.84		
14.	8,002.55		
15.	2,000.18		

Name ______________________ Date ______________________

UNIT 48 Applications

1. A bicycle was sold for $69.99 in an area with a 7% retail sales tax. What was the tax on the bike?

2. Goods with a retail price of $300 were sold in an area with a 5% retail tax. (a) What was the amount of the tax? (b) What was the total amount due from the customer?

3. If the 6% sales tax on a stove was $36, what was the price of the stove? (Hint: P = $36; R = 6%)

4. After tax, a customer paid $525 for a TV set that had a $500 retail price. What was the rate of sales tax? [Hint: $500 = B; $25 ($525 − $500 = P]

5. Goods with a list price of $600 were sold at a 20% trade discount. What was the total amount due, including sales tax, if the goods were sold in an area with a 7% sales tax rate?

ANSWERS

1. ____________
2. a. ____________
   b. ____________
3. ____________
4. ____________
5. ____________

6. On a recent trip to the supermarket, David Wong bought goods that were priced as follows: \$.69, \$2.79, \$12.99, \$5.69, \$4.99, and \$2.99. If Bill shops in an area with a 7% retail sales tax, what is (a) the total amount of the purchase and (b) the amount of sales tax?

ANSWERS

6. a. ____________

   b. ____________

UNIT 49

Finding Commissions

Instead of getting an hourly wage or a fixed salary, many salespeople work on a commission basis, that is, they get a percentage of the value of goods and services that they sell. For example, Lisa Todd is a real estate salesperson. Lisa does not get a salary; instead, she is paid 5% of the price of each house she sells. Last week, Lisa sold a house for $60,000. Her 5% commission is calculated as follows:

	$60,000	price of house
×	.05	commission rate
	$ 3,000	amount of commission

In commission problems, you are actually working with the formula **B × R = P.** The total selling price of the goods or services is B; the rate of commission is R; and P is the result of multiplying B × R. We can illustrate this with a pie diagram:

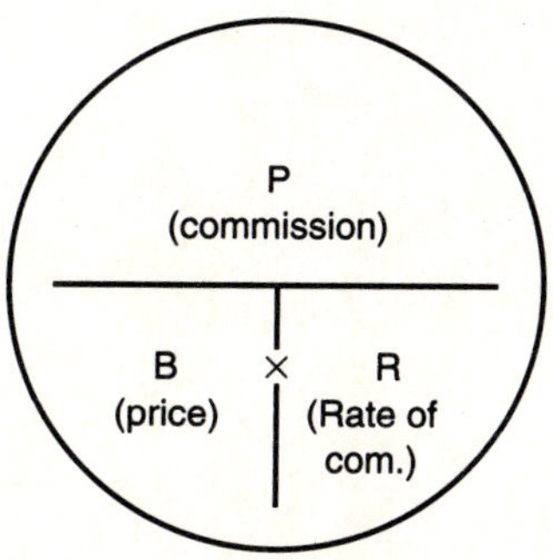

Let's look at another example.

EXAMPLE: John Distasio works for a large travel agency. John receives a 6% commission on all overseas vacations he sells. Last week, John sold a European vacation to a couple for $3,000. What was his commission?

Solution:

$$\$3{,}000 \times .06 = \$180$$

Some salespeople are paid on a salary plus commission basis. For example, Lynn Haden earns a salary of $200 a week plus a 3% commission on all merchandise she sells. Last week her total sales amounted to $4,000. Her weekly earnings are:

	$200	salary
+	120	commission ($4,000 × .03)
	$320	total earnings

Commissions are sometimes figured on a sliding scale. Let's look at a couple of examples to see how a sliding scale works.

EXAMPLE 1: Wendy Carpenter, a salesperson, gets a 4% commission on all sales up to $50,000 and a 6% commission on all sales over $50,000. Last month Wendy's total sales amounted to $62,000. What was her commission?

Solution:

Step 1: Figure commission on first $50,000:

$50,000 × .04 = $2,000

Step 2: Figure commission on sales over $50,000:

$62,000 − $50,000 = $12,000 × .06 = $720

Step 3: Add the two amounts: $2,000 + $720 = **$2,720**

EXAMPLE 2: Donald Haick gets 2% commission on the first $10,000 of goods sold each month, 3% on the next $10,000, and 5% on any goods sold in excess of $20,000. What is his commission in a month in which his sales totaled $34,000?

Solution:

$10,000 × .02 = $200

$10,000 × .03 = $300

$14,000 ($34,000 − $20,000) × .05 = $700

$200 + $300 + $700 = **$1,200 total commission**

Name ______________________________ Date ______________________________

UNIT 49 Exercises

1. The following salespeople are paid a commission based on their monthly sales. Find the commission of each.

Salesperson	Rate of Commission	Monthly Sales	Commission
John Lang	8%	$44,000	$_____
Jane Lewis	6%	$30,800	$_____
Todd King	$4\frac{1}{2}$%	$45,000	$_____
Lee Cornwell	7%	$18,000	$_____
Jesse Fazio	$8\frac{1}{2}$%	$24,000	$_____
Li Smitherman	$3\frac{3}{4}$%	$12,000	$_____
Karl Schwartz	10%	$35,600	$_____
Lou Kingrey	5%	$43,546	$_____

2. The following salespeople are paid a salary plus a commission on their weekly sales. Find the total earnings of each.

Salesperson	Salary	Rate of Commission	Weekly Sales	Total Earnings
Dave Rodgers	$225	2%	$8,000	$_____
Sueanne Lee	$340	3%	$6,200	$_____
Randy Jones	$300	5%	$3,400	$_____
Earl Hillary	$275	$4\frac{1}{2}$%	$2,500	$_____

Salesperson	Salary	Rate of Commission	Weekly Sales	Total Earnings
Lori Pope	$350	6%	$5,000	$_____
Ann Foreman	$230	7%	$6,944	$_____
Hain Bach	$500	$1\frac{1}{2}$%	$6,000	$_____
Jay Whitmore	$190	10%	$1,250	$_____

3. The following salespeople are paid a salary plus a stated commission of all weekly sales over $4,000. Find the total earnings of each.

Salesperson	Salary	Rate of Commission	Total Weekly Sales	Earnings
Taylor Dawn	$300	6%	$8,200	$_____
Alice Ingrum	$225	4%	$6,100	$_____
Henry Jordan	$300	5%	$6,500	$_____
Paul Corelli	$295	6%	$6,800	$_____

Name ______________________________ Date ______________________________

Salesperson	Salary	Rate of Commission	Total Weekly Sales	Earnings
Louis Crowell	$235	4%	$5,600	$_____
Wendy Cars	$325	5%	$4,900	$_____
Lisa Dobbs	$345	7%	$5,675	$_____

Work Space

Name ______________________ Date ______________________

UNIT 49 Applications

ANSWERS

1. ____________
2. ____________
3. ____________
4. ____________

1. Tim Reed is a salesperson for Harvey Lumber Company. Tim gets a salary of \$325 weekly plus a 4% commission on his total weekly sales. This week his total sales amounted to \$9,200. What is his total earnings?

2. Renee Friedman receives a monthly salary of \$1,200 plus a 6% commission on monthly sales in excess of \$12,000. Last month her sales totaled \$14,700. What is her total earnings?

3. Randy Knight works on a straight commission basis. His commission is paid on a sliding scale, as follows:

   2% on first \$10,000 sold

   3% on next \$15,000 sold

   5% on next \$15,000 sold

   8% on all sales in excess of \$40,000

   Last month Randy's total sales amounted to \$47,500. What is his commission?

4. Joanne Preston's 6% commission for selling a house came to \$3,600. What was the selling price of the house? (Hint: R = 6%, P = \$3,600, find B)

5. Pylant Products Company pays its salespeople a base salary and a commission based on the following sliding scale: 5% of first $1,000, 8% of next $1,000, and 10% of all sales above $2,000. Complete the following earnings report.

Pylant Products Company
Monthly Earnings Report
For Month Ended July 31, 19--

Salesperson	Total Sales	Amount of Commission	Base Salary	Total Earnings
Lynn Porter	$4,800	$______	$1,000	$______
Jan Herdanez	$2,450	$______	$1,200	$______
Dave Stover	$3,500	$______	$1,490	$______
Kay Segram	$2,300	$______	$1,450	$______
Lyle Kirkland	$4,560	$______	$1,250	$______
Anne Moore	$6,700	$______	$1,425	$______
Vickie Cooke	$4,200	$______	$1,230	$______
Totals		$______		$______

UNIT 50

Interest

The use of credit plays a very vital role in our nation's economy. Each business day countless numbers of businesses and individuals gain immediate possession of goods and services with the understanding that payment for these goods and services will be made in the future. Users of credit range from a high school student buying a $30 sweater on credit at a local department store to Donald Trump's $300,000,000 credit purchase of the airline that is now called Trump Shuttle.

THE CHARGE FOR CREDIT

When you borrow money, or make a credit purchase, you will usually have to pay a fee for the use of credit. This fee is called **interest; it is money paid for the use of money.** The amount of interest you will be charged depends on three factors:

1. **Principal.** The principal is the amount borrowed.
2. **Rate.** The rate is the percent of interest being charged.
3. **Time.** The time is the length of time the debt will be outstanding.

To illustrate these factors, let's take an example. Bill Hardy wants to open his own carpet cleaning business. Bill estimates that he can get started in his new business with a $10,000 loan. So, Bill visited his local bank to talk with a loan officer. Bill found out that if he was willing to use the value of his house as security for the loan, the bank was willing to lend him the $10,000 he needs. Bill accepted these terms and agreed to repay the loan in one year, with interest at 10%.

The *principal* of Bill's loan is $10,000, the amount he borrowed. The *rate* of Bill's loan is 10%, the percent the bank charged. And the *time* of Bill's loan is one year, the length of time before Bill has to repay his loan.

INTEREST FOR A FULL YEAR

As stated above, interest is expressed as a rate (remember, "rate" means the same thing as "percent"). And, unless stated otherwise, the rate expressed is an *annual* rate. Thus, a rate of 10% means *10% of the principal for one year.* Therefore, to find the interest for one year, just multiply the principal by the rate.

EXAMPLE: How much interest will Bill Hardy have to pay for the use of $10,000 for one year at 10%?
Solution:

Principal	×	Rate	=	Interest for One Year
$10,000	×	.10	=	**$1,000**

THE INTEREST FORMULA

To find the interest for periods other than one year, we usually use the **interest formula,** which is:

$$\text{Interest} = \text{Principal} \times \text{Rate} \times \text{Time} \quad \text{or}$$
$$I = P \times R \times T$$

To illustrate, let's find the interest on a $10,000 loan at 10% interest for 2 years:

$$P \times R \times T$$
$$\$10{,}000 \times .10 \times 2 = \mathbf{\$2{,}000}$$

Let's look at a few more examples.

EXAMPLE 1: Find the interest on a $12,000 loan at 12% interest for 3 years.
Solution:

$$P \times R \times T$$
$$\$12{,}000 \times .12 \times 3 = \mathbf{\$4{,}320}$$

EXAMPLE 2: Find the interest on an $1,850 credit purchase of a sound system financed for 3 years at $12\frac{1}{2}\%$ interest.
Solution:

$$P \times R \times T$$
$$\$1{,}850 \times .125 \times 3 = \mathbf{\$693.75}$$

Remember: $12\frac{1}{2}\% = 12.5\% = 12.5\% = .125$

INTEREST FOR FRACTIONS OF A YEAR

Credit periods will very often be for periods of time that are less than a year. For example, 30 days, 60 days, 90 days, 6 months, 9 months, and so on. When this happens, you state the *T* (time) in the interest formula as a fractional part of a year.

Time Expressed in Months. If the time of a loan is expressed in months, the number of months is placed over 12 (12 months in one year). So, 5 months becomes $\frac{5}{12}$; 3 months becomes $\frac{3}{12}$ or $\frac{1}{4}$; and 6 months becomes $\frac{6}{12}$ or $\frac{1}{2}$. To illustrate, let's calculate the interest on a $1,200 loan at 12% interest for 3 months:

$$P \times R \times T$$
$$\$1{,}200 \times .12 \times \frac{1}{4}$$
$$= \$144 \times \frac{1}{4}$$
$$= \mathbf{\$36}$$

Remember: $\frac{3}{12} = \frac{1}{4}$

To solve this problem, we multiplied P by R ($1,200 × .12 = $144). Then we multiplied $144 by $\frac{1}{4}$ to get the interest charge. It is often easier to calculate interest by using cancellation. For example, we can use cancellation to calculate interest on the above problem as follows:

$$\frac{\cancel{\$1{,}200}^{\,12}}{1} \times \frac{\cancel{12}^{\,3}}{\cancel{100}_{\,1}} \times \frac{1}{\cancel{4}_{\,1}} = \$36$$

The rate is expressed as $\frac{12}{100}$ because percent means by the hundred.

Times Expressed in Days. To simplify the computation of interest when the time is expressed in days, the business community usually uses a 360-day year (called the **banker's year**) instead of the actual year of 365 days. Use of the 365-day year will yield **exact interest** and is used by the U.S. government. It is, however, used infrequently in business. Let's look at a couple of examples of interest calculation using a 360-day year.

EXAMPLE 1: Find the interest on a $300, 90-day loan at a rate of 10%.
Solution:

$$\frac{\cancel{\$300}^{\,3}}{1} \times \frac{\cancel{10}^{\,5}}{\cancel{100}_{\,1}} \times \frac{\cancel{90}^{\,1}}{\cancel{360}_{\,\cancel{4}_{\,2}}} = \$7.50$$

EXAMPLE 2: Find the interest on an $800 loan for 60 days at 9% interest.
Solution:

$$\frac{\cancel{\$800}^{\,\cancel{8}^{\,4}}}{1} \times \frac{\cancel{9}^{\,3}}{\cancel{100}_{\,1}} \times \frac{\cancel{60}^{\,1}}{\cancel{360}_{\,\cancel{6}_{\,\cancel{2}_{\,1}}}} = \$12$$

Name ____________________ Date ____________________

UNIT 50 Exercises

Find the interest in each of the following.

	Principal	Rate	Time	Interest
1.	$1,200	8%	1 year	_____
2.	$2,600	12%	3 years	_____
3.	$ 200	14%	4 months	_____
4.	$1,250	10%	9 months	_____
5.	$4,500	11%	60 days	_____
6.	$ 800	$12\frac{1}{2}\%$	30 days	_____

	Principal	Rate	Time	Interest
7.	$ 2,890	10%	45 days	______
8.	$12,000	15%	180 days	______
9.	$45,000	9%	120 days	______
10.	$62,500	12%	240 days	______
11.	$14,000	8%	90 days	______
12.	$95,000	14%	70 days	______

Name ______________________ Date ______________________

UNIT 50 Applications

1. What will be the interest charge on a $5,400 loan for 180 days at an interest rate of 9%?

2. Furgerson Wholesale Grocery Company needs to borrow $45,000 for 6 months. What will be the interest charge if a local bank is willing to make the loan at a rate of 12%?

3. To pay various bills, Don Myers borrowed $2,000 for 2 years at an interest rate of 10%. What will be his interest charge?

4. To take advantage of a special purchase price on word processors, Gary Lyons needs to borrow $3,560 for 30 days. What will be the interest charge if Gary can borrow the money at 9%?

5. Larry Swift is just starting a new law practice. The cost of law books, furniture, and equipment he needs is $18,000. What will be the interest charge if Bill can borrow this amount for 2 years at 10% interest?

ANSWERS

1. ____________
2. ____________
3. ____________
4. ____________
5. ____________

6. Harkon Building Company borrowed the following amounts during the last year:

$25,000 for 120 days at 12%.

$38,000 for 2 months at 15%.

$125,000 for 9 months at 11%.

What was the total amount Harkon paid in interest during the year?

ANSWERS

6. ____________

UNIT 51 Determining Gross Pay

The total amount an employee earns during a payroll period is called **gross pay** or **gross earnings.** Gross pay can be determined in a variety of ways. In Unit 49, for example, we looked at various commission plans. In this unit, we are going to look at the two most common methods of determining gross pay: (1) an **hourly wage** (or **hourly rate**) and (2) a fixed **salary.** Both methods are discussed next.

HOURLY WAGES

Most employees who work for an hourly wage are covered by the **Fair Labor Standards Act,** which is a federal law that sets the minimum hourly wage and the standard work week. At this writing, the *minimum* hourly wage is \$4.25 for a standard work week of 40 hours. If an employee works more than 40 hours in a given work week, the employee must be paid at least one and one-half times the regular hourly rate for all hours worked in excess of 40. This is called **overtime.** We will look at overtime shortly. For the moment, let's look at an example of finding gross pay when overtime does not apply.

EXAMPLE: Joy Bodiford earns an hourly wage of \$5.00. Last week Joy worked 40 hours. What was her gross pay?
Solution:

Hours worked	×	Rate per hour	=	Gross pay
40	×	\$5.00	=	\$200.00

In this example, Joy earns \$5.00 an hour, which is more than the minimum wage. Remember the key word here is *minimum.* Many companies pay well in excess of the minimum wage as a matter of company policy. And some companies are not required to pay the minimum wage—generally, these are companies that are not involved in interstate commerce.

Let's now look at an example where an employee works more than 40 hours in a week. Remember, the law requires the employer to pay at least one and one-half times the regular hourly rate for overtime. Thus, the overtime rate can be found by multiplying the regular hourly rate by 1.5.

EXAMPLE: Clayton Morrow earns an hourly wage of \$6.00. Last week he worked a total of 44 hours. What is his gross pay?
Solution:

Step 1: Find his regular pay for 40 hours:

Regular hours	×	Hourly wage	=	Regular pay
40	×	\$6.00	=	\$240

Step 2: Find his overtime pay:

Hours over 40	×	Overtime rate	=	Overtime pay
4	×	\$9 (\$6 × 1.5)	=	\$36

Step 3: Add regular pay and overtime pay to get gross pay:

\$240 + \$36 = **\$276**

Since overtime pay is at least one and one-half times the regular rate of pay, it is often called **time and a half.**

Employers sometimes calculate overtime at a rate other than 1.5 times the regular hourly rate. For example, some companies pay overtime if total hours worked in a day exceed 8, even though total hours for the week do not exceed 40. Additionally, some companies pay double time (and sometimes triple time) for work on weekends and holidays. But it is important to remember that such plans are a matter of individual company policy, not the law. Let's look at an example of an employee who earns double time for any work on Saturday.

EXAMPLE: Reggie South earns a regular hourly wage of $8.00. Reggie gets overtime for any hours worked over 40, and double time for any hours worked on Saturday. Last week Reggie worked 48 hours, 4 of which were on Saturday. What is his gross pay?

Solution:

Step 1: Find his regular pay for 40 hours:

$$40 \times \$8 = \$320$$

Step 2: Find his overtime pay (4 regular overtime hours):

$$4 \times \$12\ (\$8 \times 1.5) = \$48$$

Step 3: Find his double time pay (4 hours on Saturday):

$$4 \times \$16\ (\$8 \times 2) = \$64.$$

Step 4: Add the three amounts to find his gross pay:

$$\$320 + \$48 + \$64 = \$432$$

SALARIES

Salary is usually stated as a certain amount of money paid to an employee on an annual basis. Salaried employees are usually paid the same, whether hours worked are 40 and under, or over 40. You find the amount of a salaried person's paycheck by dividing the annual salary by the number of pay periods in the year. Pay periods for salaried people are usually on a **weekly, biweekly, semimonthly,** or **monthly** basis. To illustrate, let's say that an employee is hired at an annual salary of $16,900. We can calculate gross pay per period as follows:

Pay Period	Number of Pay Periods in a Year	Gross Pay Per Pay Period
Weekly	52	$ 325.00 ($16,900 ÷ 52)
Biweekly	26	650.00 ($16,900 ÷ 26)
Semimonthly	24	704.17 ($16,900 ÷ 24)
Monthly	12	1,408.33 ($16,900 ÷ 12)

Name ____________________ Date ____________________

► UNIT 51 Exercises

1. Calculate regular pay and gross pay for each of the following employees. All employees are paid time and a half if hours exceed 40.

Employee	Hourly Pay	Hours Worked	Regular Pay	Total Pay
A	$8.00	44	$______	$______
B	$6.25	45	$______	$______
C	$4.00	25	$______	$______
D	$5.00	50	$______	$______
E	$7.50	48	$______	$______
F	$6.90	52	$______	$______

Employee	Hourly Pay	Hours Worked	Regular Pay	Total Pay
G	$7.40	40	$______	$______
H	$4.75	46	$______	$______
I	$9.50	54	$______	$______
J	$8.50	47	$______	$______

Name ______________________________ Date ______________________________

2. Find the gross pay of each of the following employees. All employees are paid time and a half for hours over 40 and double time for Sunday work.

Employee	Hours Worked							Total Hours	Hourly Rate	Regular Pay	Overtime Pay	Gross Pay
	M	T	W	T	F	S	S					
A. Smith	8	8	7	8	8	7		______	$ 6.50	$______	$______	$______
B. Kaye	8	8	8	8	8		4	______	8.25	______	______	______
S. Snead	7	6	9	9	9	2		______	6.00	______	______	______
T. Lee	8	8	8	8	8			______	12.00	______	______	______
H. Dobbs	8	8	8	8	8	4	6	______	8.00	______	______	______
K. Frost	8	8	8	9	8			______	10.00	______	______	______

Work Space

Name ______________________ Date ______________________

UNIT 51 Applications

ANSWERS

1. ____________
2. ____________
3. ____________
4. ____________
5. ____________

1. James Sanders makes $8.25 an hour and his employer is covered by the Fair Labor Standards Act. Last week, James worked a total of 49 hours. What is his gross pay?

2. Sue Chappell worked 34 hours in one week. What is her gross pay if her hourly wage is $4.75?

3. Last week Lin Todd worked 40 hours at an hourly wage of $8.00. What is his gross pay?

4. Kathy James is paid 1.5 times her regular pay for all hours worked over 40 in a week. Last week Kathy worked 45 hours and is paid a regular hourly wage of $9.00. What was her gross pay for the week?

5. Patricia Belser has a salaried job that pays her $450 a week. Last week Patricia worked 42 hours. What is her gross pay?

6. Rob Abrams was just hired at an annual salary of $18,720. What is Rob's gross pay if his pay period is:

   a. Weekly
   b. Biweekly
   c. Semimonthly
   d. Monthly

7. Tammy Lewis gets paid 1.5 times her regular hourly rate if she works more than 8 hours in a day, even though she may not work more than 40 hours in the week. Last week she worked 8 hours on Monday, 10 hours on Tuesday, 10 hours on Wednesday, and 6 hours on Thursday. What is her gross pay if her regular hourly wage is $6.50?

8. Sean Wilson earns $9.00 an hour and he gets time and a half for hours over 40 and double time for work on holidays. Last week he worked 54 hours, 8 of which were on a holiday. What is his gross pay?

ANSWERS

6. a. ________
   b. ________
   c. ________
   d. ________
7. ________
8. ________

UNIT 52 Payroll Deductions and the Payroll Register

In Unit 50, we learned how to calculate gross pay for employees who work for a salary and employees who work for an hourly wage. In this unit, we are going to study how to find **net pay.** Net pay (also called **take-home pay**) is the amount left after all payroll deductions have been made; it is the actual amount of an employee's paycheck.

Payroll deductions can be grouped into two categories: (1) those required by law and (2) those agreed to by the employee and the employer. Let's look at those required by law first.

PAYROLL DEDUCTIONS REQUIRED BY LAW

The federal government and most state governments require employers to withhold income taxes from the earnings of employees. The federal government also requires employers to withhold FICA (social security) from the wages of employees.

Federal Income Tax. The main source of revenue for our country comes from the income tax imposed on individual employees. The amount of federal income tax an employee pays depends on three factors: (1) the employee's gross pay, (2) the employee's marital status, and (3) the number of **withholding allowances** claimed by the employee.

We calculated gross pay in Unit 50. An employee's marital status, for the purpose of withholding income tax, is either married or single. A withholding allowance, also called an **exemption,** is an amount of earnings not subject to income taxes. An exemption is allowed for the employee and each person for whom the employee provides support.

To get the actual amount of federal income tax to be withheld, most employers use tax tables published by the Internal Revenue Service (an agency of the federal government). Tax tables are part of a publication called **Circular E,** which provides tax tables for weekly, biweekly, semimonthly, monthly, and daily or miscellaneous payroll periods for married and single persons. Tax tables for *Single Persons—Weekly Payroll Period* and *Married Persons—Weekly Payroll Period* are shown on pages 444 and 445.

To illustrate the use of tax tables, we will use Lang Kirkland as our example. Lang is married and claims three withholding allowances (one for himself, one for his wife, and one for his child). Lang works by the week and last week his gross pay was $350. To figure Lang's federal income tax, look at the Married Persons—Weekly Payroll table shown on page 445. Lang's pay, $350, is in the bracket of "at least $350 but less than $360." So, with three withholding allowances, Lang's federal income tax is $28. Find this amount in the table.

State and Local Income Taxes. As stated earlier, most state governments (and also some city governments) require employers to withhold an income tax from the earnings of employees. The rates for these taxes vary so greatly from state to state that it would not be possible to show all the various tax tables here. But, like the federal government, state governments

Single Persons—Weekly Payroll Period												
And the wages are—		And the number of withholding allowances claimed is—										
At least	But less than	0	1	2	3	4	5	6	7	8	9	10
		The amount of income tax to be withheld shall be—										
$ 0	$ 25	$ 0	$ 0	$ 0	$ 0	$ 0	$ 0	$ 0	$ 0	$ 0	$0	$0
25	30	1	0	0	0	0	0	0	0	0	0	0
30	35	2	0	0	0	0	0	0	0	0	0	0
35	40	3	0	0	0	0	0	0	0	0	0	0
40	45	3	0	0	0	0	0	0	0	0	0	0
45	50	4	0	0	0	0	0	0	0	0	0	0
50	55	5	0	0	0	0	0	0	0	0	0	0
55	60	6	0	0	0	0	0	0	0	0	0	0
60	65	6	1	0	0	0	0	0	0	0	0	0
65	70	7	1	0	0	0	0	0	0	0	0	0
70	75	8	2	0	0	0	0	0	0	0	0	0
75	80	9	3	0	0	0	0	0	0	0	0	0
80	85	9	4	0	0	0	0	0	0	0	0	0
85	90	10	4	0	0	0	0	0	0	0	0	0
90	95	11	5	0	0	0	0	0	0	0	0	0
95	100	12	6	0	0	0	0	0	0	0	0	0
100	105	12	7	1	0	0	0	0	0	0	0	0
105	110	13	7	2	0	0	0	0	0	0	0	0
110	115	14	8	3	0	0	0	0	0	0	0	0
115	120	15	9	3	0	0	0	0	0	0	0	0
120	125	15	10	4	0	0	0	0	0	0	0	0
125	130	16	10	5	0	0	0	0	0	0	0	0
130	135	17	11	6	0	0	0	0	0	0	0	0
135	140	18	12	6	1	0	0	0	0	0	0	0
140	145	18	13	7	1	0	0	0	0	0	0	0
145	150	19	13	8	2	0	0	0	0	0	0	0
150	155	20	14	9	3	0	0	0	0	0	0	0
155	160	21	15	9	4	0	0	0	0	0	0	0
160	165	21	16	10	4	0	0	0	0	0	0	0
165	170	22	16	11	5	0	0	0	0	0	0	0
170	175	23	17	12	6	0	0	0	0	0	0	0
175	180	24	18	12	7	1	0	0	0	0	0	0
180	185	24	19	13	7	2	0	0	0	0	0	0
185	190	25	19	14	8	3	0	0	0	0	0	0
190	195	26	20	15	9	3	0	0	0	0	0	0
195	200	27	21	15	10	4	0	0	0	0	0	0
200	210	28	22	16	11	5	0	0	0	0	0	0
210	220	29	24	18	12	7	1	0	0	0	0	0
220	230	31	25	19	14	8	3	0	0	0	0	0
230	240	32	27	21	15	10	4	0	0	0	0	0
240	250	34	28	22	17	11	6	0	0	0	0	0
250	260	35	30	24	18	13	7	1	0	0	0	0
260	270	37	31	25	20	14	9	3	0	0	0	0
270	280	38	33	27	21	16	10	4	0	0	0	0
280	290	40	34	28	23	17	12	6	0	0	0	0
290	300	41	36	30	24	19	13	7	2	0	0	0
300	310	43	37	31	26	20	15	9	3	0	0	0
310	320	44	39	33	27	22	16	10	5	0	0	0
320	330	46	40	34	29	23	18	12	6	1	0	0
330	340	47	42	36	30	25	19	13	8	2	0	0
340	350	49	43	37	32	26	21	15	9	4	0	0
350	360	50	45	39	33	28	22	16	11	5	0	0
360	370	52	46	40	35	29	24	18	12	7	1	0
370	380	55	48	42	36	31	25	19	14	8	3	0
380	390	58	49	43	38	32	27	21	15	10	4	0

Married Persons—Weekly Payroll Period												
And the wages are—		And the number of withholding allowances claimed is—										
At least	But less than	0	1	2	3	4	5	6	7	8	9	10
		The amount of income tax to be withheld shall be—										
180	185	19	13	7	2	0	0	0	0	0	0	0
185	190	19	14	8	2	0	0	0	0	0	0	0
190	195	20	14	9	3	0	0	0	0	0	0	0
195	200	21	15	10	4	0	0	0	0	0	0	0
200	210	22	16	11	5	0	0	0	0	0	0	0
210	220	23	18	12	7	1	0	0	0	0	0	0
220	230	25	19	14	8	2	0	0	0	0	0	0
230	240	26	21	15	10	4	0	0	0	0	0	0
240	250	28	22	17	11	5	0	0	0	0	0	0
250	260	29	24	18	13	7	1	0	0	0	0	0
260	270	31	25	20	14	8	3	0	0	0	0	0
270	280	32	27	21	16	10	4	0	0	0	0	0
280	290	34	28	23	17	11	6	0	0	0	0	0
290	300	35	30	24	19	13	7	2	0	0	0	0
300	310	37	31	26	20	14	9	3	0	0	0	0
310	320	38	33	27	22	16	10	5	0	0	0	0
320	330	40	34	29	23	17	12	6	1	0	0	0
330	340	41	36	30	25	19	13	8	2	0	0	0
340	350	43	37	32	26	20	15	9	4	0	0	0
350	360	44	39	33	28	22	16	11	5	0	0	0
360	370	46	40	35	29	23	18	12	7	1	0	0
370	380	47	42	36	31	25	19	14	8	2	0	0
380	390	49	43	38	32	26	21	15	10	4	0	0
390	400	50	45	39	34	28	22	17	11	5	0	0
400	410	52	46	41	35	29	24	18	13	7	1	0
410	420	53	48	42	37	31	25	20	14	8	3	0
420	430	55	49	44	38	32	27	21	16	10	4	0
430	440	56	51	45	40	34	28	23	17	11	6	0
440	450	58	52	47	41	35	30	24	19	13	7	2
450	460	59	54	48	43	37	31	26	20	14	9	3
460	470	61	55	50	44	38	33	27	22	16	10	5
470	480	62	57	51	46	40	34	29	23	17	12	6
480	490	64	58	53	47	41	36	30	25	19	13	8
490	500	65	60	54	49	43	37	32	26	20	15	9
500	510	67	61	56	50	44	39	33	28	22	16	11
510	520	68	63	57	52	46	40	35	29	23	18	12
520	530	70	64	59	53	47	42	36	31	25	19	14
530	540	71	66	60	55	49	43	38	32	26	21	15
540	550	73	67	62	56	50	45	39	34	28	22	17
550	560	74	69	63	58	52	46	41	35	29	24	18

provide tax tables and, also like the federal government, base the tax on an employee's gross pay, marital status, and withholding allowances.

For our example of Lang Kirkland, we will assume that his state income tax for this pay period is $12.

FICA Taxes. The **Federal Insurance Contributions Act (FICA),** more commonly called **social security,** requires employers to withhold a certain percentage of each employee's earnings as a contribution to the federal social security program.

There are two parts to the FICA tax: (1) **OASDI** (old age, survivors, and disability insurance) and (2) **medicare** (hospital insurance). Before 1991, employers could combine and report the FICA tax as a single amount. Starting in 1991, however, the two parts of the tax are calculated on different bases. For OASDI, the current rate is 6.2% of the first $53,400 earned in the year. For medicare, the current rate is 1.45% of the first $125,000 earned in the year. If an employee's earnings were to exceed these bases ($53,400 for OASDI and $125,000 for medicare), he or she would pay no more FICA taxes for the rest of the year.

Now, let's return to our example of Lang Kirkland. Lang's weekly earnings were $350 and he has not reached $53,400 in earnings for the year. So, his FICA tax for this payroll period is:

OASDI	$350 × .062 =	$21.70
Medicare	350 × .0145 =	5.08
Total		$26.78

DEDUCTIONS AGREED TO BY THE EMPLOYEE AND THE EMPLOYER

We have discussed the most common deductions that are required by law to be withheld. There are many other deductions that an employee can agree to have withheld from his or her pay. The more common are health insurance, retirement plans, savings plans, loan repayments, union dues, and amounts to purchase U.S. Government savings bonds. For our example of Lang Kirkland, we will assume that he has $25 withheld weekly for savings bonds.

CALCULATING NET PAY

As stated earlier, net pay is the amount of an employee's check after all deductions have been made. Lang Kirkland's net pay is calculated as follows:

Gross pay		$350.00
Less: Federal income tax	$28.00	
State income tax	12.00	
FICA tax	26.78	
Savings bonds	25.00	
Total deductions		91.78
Net pay		$258.22

THE PAYROLL REGISTER

The **payroll register** is a form used by businesses to organize and summarize payroll data. The actual design of the payroll register depends on the needs of the individual company. The payroll register shown on page 447 is for Allied Products Company, a distributor of soft drinks.

PAYROLL REGISTER

FOR WEEK ENDING June 29 **19** XX

	Employee's Name	Marital Status	No. W.H. Allow.	Regular Earnings: Hours Worked	Regular Earnings: Rate per Hour	Regular Earnings: Amount	Overtime Earnings: Hours Worked	Overtime Earnings: Rate per Hour	Overtime Earnings: Amount	Total Earnings	Deductions: FICA Tax	Deductions: Federal Income Tax	Deductions: State Income Tax	Deductions: Other	Net Pay
1	Adams, Phillip H.	M	3	40	8 00	320 00	—	—	—	320 00	24 48	23 00	14 20	4 00	254 32
2	Barker, Patrice R.	M	1	40	9 00	360 00	—	—	—	360 00	27 54	40 00	18 60	4 00	269 86
3	Dansky, Randolph P.	M	4	44	10 50	420 00	4	15 75	63 00	483 00	36 95	41 00	21 00	—	384 05
4	Franklin, Pamela R.	M	1	40	6 00	240 00	—	—	—	240 00	18 36	22 00	8 70	—	190 94
5	Helms, Faye	M	2	40	12 00	480 00	—	—	—	480 00	36 72	53 00	23 40	4	362 88
6	Purdue, Joy H.	M	2	46	6 00	240 00	6	9 00	54 00	294 00	22 49	24 00	12 00	15 00	220 51
7	Turner, Joseph J.	M	1	38	8 00	304 00	—	—	—	304 00	23 26	31 00	14 00	4 00	231 74
8	Veal, Michele	S	1	40	8 00	320 00	—	—	—	320 00	24 48	40 00	18 02	—	237 50
9	Watson, Teresa L.	M	1	40	6 00	240 00	—	—	—	240 00	18 36	22 00	8 70	4 00	186 94
10	Williams, Todd K.	S	1	40	9 00	360 00	—	—	—	360 00	27 54	46 00	22 90	—	263 56
11	Totals					3284 00				3401 00	260 18	342 00	161 52	35 00	2602 30
12															
13															
14															
15															
16															
17															
18															
19															
20															
21															

Work Space

Name ______________________ Date ______________________

UNIT 52 Exercises

1. Using the tax tables shown in this unit, calculate the amount of federal income tax to be withheld from the earnings of each of the following employees.

Employee	Marital Status	Withholding Allowances	Gross Pay	Amount of Income Tax
A. Dye	M	4	$380	$_____
S. Kaye	S	1	$325	$_____
B. Birch	M	2	$402	$_____
M. Man	S	1	$290	$_____
H. Book	M	5	$425	$_____
C. Hendrix	M	1	$425	$_____
M. Mallard	M	3	$395	$_____
T. O'Neal	S	1	$389	$_____

2. Calculate the federal income tax, state income tax, FICA tax, and net pay for each of the following employees. Assume a rate of 8% for state income tax. No employee has reached or exceeded $53,400 in earnings.

Employee	Marital Status	Withhold. Allowance	Gross Pay	Federal Income Tax	State Income Tax	Total FICA	Net Pay
L. Walk	M	4	$480	$______	$______	$______	$______
W. Mays	S	1	$340	$______	$______	$______	$______
B. Little	M	5	$390	$______	$______	$______	$______
D. Kramer	M	2	$422	$______	$______	$______	$______
E. Hall	S	1	$372	$______	$______	$______	$______

3. Calculate the FICA tax on each of the following employees.

Employee	Earnings Before This Pay Period	Earnings This Pay Period	FICA	
			OASDI	Medicare
A	$12,800	$450	$______	$______
B	$53,100	$800	$______	$______
C	$56,000	$980	$______	$______
D	$34,500	$535	$______	$______
E	$53,050	$590	$______	$______

Name ______________________ Date ______________________

UNIT 52 Applications

1. James Jordan is married and claims three withholding allowances. His weekly salary is $320. What is the amount of federal income tax to be withheld?

2. Sue Lesan is single and claims one withholding allowance. Last week her gross pay was $325. Compute her federal income tax, FICA tax, and net pay. Assume her state income tax is $12.

3. Glen Dyer's gross pay for the current pay period was $840. What is his OASDI tax if his earnings before this pay period were (a) $53,100, (b) $41,900?

4. Danny Czonka is married and claims four withholding allowances. Last week his gross pay was $390 and his total earnings for the year are $21,500. Find (a) his federal income tax, (b) FICA tax, and (c) net pay, assuming he has state income tax of $14 and health insurance deduction of $24.

ANSWERS

1. ________
2. ________
3. a. ________
   b. ________
4. a. ________
   b. ________
   c. ________

5. Linda Day's take-home pay for this week was $220. Her gross pay was $300. What percent of her gross was withheld? [Hint: P = $80 ($300 − $220), B = $300, Find R]

ANSWERS

5. ____________

6. ____________

6. Tom Posney's FICA tax for the week was $17.21, which was 7.65% of his gross pay. What was his gross pay? (Hint: P = $17.21, R = .0765, find B).

7. The partial payroll register of Goozner Company is shown on page 453. Complete the payroll register.

PAYROLL REGISTER

FOR WEEK ENDING AUGUST 15 **19** XX

	Employee's Name	Marital Status	No. W.H. Allow.	Regular Earnings: Hours Worked	Regular Earnings: Rate Per Hour	Regular Earnings: Amount	Overtime Earnings: Hours Worked	Overtime Earnings: Rate Per Hour	Overtime Earnings: Amount	Total Earnings	Deductions: FICA Tax	Deductions: Federal Income Tax	Deductions: State Income Tax	Deductions: Other	Net Pay
1	ABRAHMS, POLLY	M	2	40	9 00		2						8 00	6 00	
2	ADAMS, RAY	M	4	40	10 00		8						10 40	—	
3	CONNERS, TIMOTHY	M	1	40	8 00		—						6 40	—	
4	DENTON, ROBERT	S	0	40	7 00		4						6 44	—	
5	GREY, TAYLOR	M	2	40	11 00		—						8 80	6 00	
6	JONES, KAYE	S	1	40	9 50		—						7 60	—	
7	KITCHENS, WYLENE	M	1	40	11 00		—						8 90	—	
8	MAYS, WANDA	M	4	40	12 00		—						9 60	—	
9	PATTERSON, LANE	S	1	40	9 00		—						7 20	—	
10	ROBERTS, FRED	M	2	40	10 00		8						10 08	6 00	
11	TOTALS												83 42	18 00	
12															
13															
14															
15															
16															
17															
18															
19															
20															
21															

Work Space

UNIT
53

Bank Checking Account

A bank is a financial institution that offers various services to its customers. One of the most common and most important services offered by banks is the **checking account.** A checking account is an amount of cash that a person or business has **on deposit** with a bank. The use of bank checking accounts, for personal use as well as business use, has become a near universal business practice. Not only does a checking account provide convenience in making cash payments, it provides control and protection of cash.

BANK CHECKS

A check is a written order directing a bank to pay a specified sum of money to a specified person or business. A check is said to be **drawn** against the account of the person writing the check. Thus, a person or business who writes a check is called the **drawer.** The person or business the check is made payable to is called the **payee.** And the bank honoring the check for payment is called the **drawee.**

Checks vary in form, depending on the needs of the user. A common form of check has a stub attached for recording the date, amount, and purpose of the check. This type of check looks like the one below.

CHECK WRITING PROCEDURE

Errors that can be costly in terms of time and money can be minimized—or avoided completely—by following established procedures when writing checks. For best results, these procedures should be followed:

1. The check stub should be filled out first; otherwise this important form could be overlooked and a check could be mailed without a record.
2. The same date entered on the check stub should be entered on the **Date** line of the check.
3. The name of the payee should be clearly written on the **Pay to the Order Of** line.

4. The amount of the check should be written numerically after the printed dollar sign (on the same line with the name of the payee). The numbers should be written as close as possible to the dollar sign to eliminate the possibility of alteration.

5. The amount of the check should be written out in words on the line directly under the **Pay to the Order Of** line. Dollars are written out in words, and cents are written in terms of hundredths of a dollar. A line should be drawn to fill in empty space.

6. The check should be signed by an authorized person.

7. Corrections should not be made on a check. Should a mistake be made, write the word **VOID** on both the check and the check stub and write a new check.

To illustrate the check writing process, let's look at the example below, which shows several checks written by Thomas B. Anderson during June 19XX.

MAKING BANK DEPOSITS

To minimize the amount of cash on hand, and thus minimize the risk of loss, most businesses make daily deposits of cash and checks received. A **deposit slip** or **deposit ticket** should be prepared each time a deposit is made. A completed deposit slip is shown on p. 457. Notice that space is provided for listing currency (paper money), coins, and checks that are being deposited.

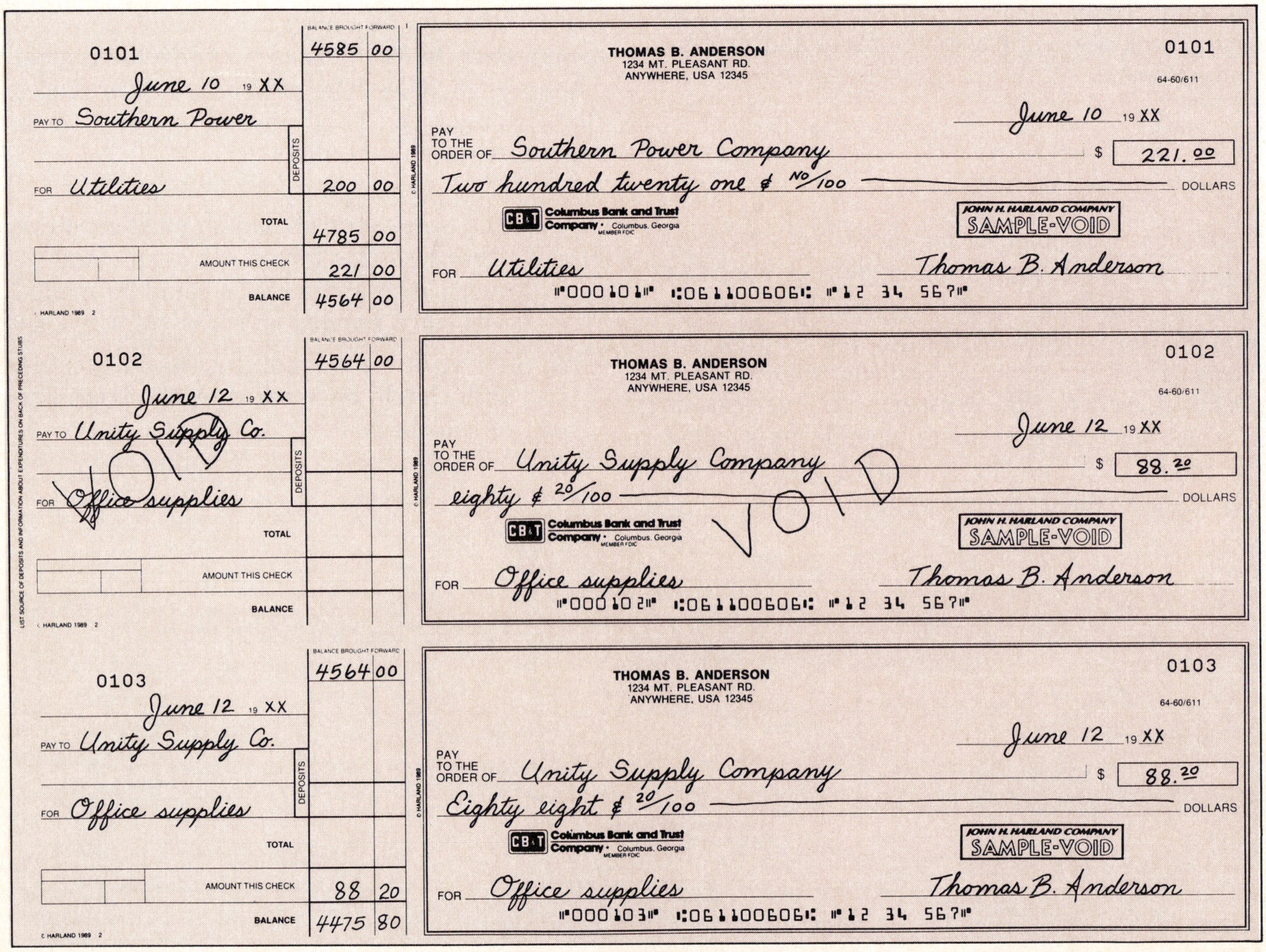

0101 — June 10 19 XX — PAY TO Southern Power — FOR Utilities

BALANCE BROUGHT FORWARD 4585 00 — DEPOSITS 200 00 — TOTAL 4785 00 — AMOUNT THIS CHECK 221 00 — BALANCE 4564 00

THOMAS B. ANDERSON
1234 MT. PLEASANT RD.
ANYWHERE, USA 12345
0101
64-60/611
June 10 19 XX
PAY TO THE ORDER OF Southern Power Company $ 221.00
Two hundred twenty one & NO/100 DOLLARS
Columbus Bank and Trust Company • Columbus, Georgia MEMBER FDIC
JOHN H. HARLAND COMPANY SAMPLE-VOID
FOR Utilities Thomas B. Anderson
⑈000101⑈ ⑆061100606⑆ ⑈12 34 567⑈

0102 — June 12 19 XX — PAY TO Unity Supply Co. — FOR Office supplies — VOID

BALANCE BROUGHT FORWARD 4564 00 — DEPOSITS — TOTAL — AMOUNT THIS CHECK — BALANCE

THOMAS B. ANDERSON
1234 MT. PLEASANT RD.
ANYWHERE, USA 12345
0102
64-60/611
June 12 19 XX
PAY TO THE ORDER OF Unity Supply Company $ 88.20
eighty & 20/100 DOLLARS
VOID
Columbus Bank and Trust Company • Columbus, Georgia MEMBER FDIC
JOHN H. HARLAND COMPANY SAMPLE-VOID
FOR Office supplies Thomas B. Anderson
⑈000102⑈ ⑆061100606⑆ ⑈12 34 567⑈

0103 — June 12 19 XX — PAY TO Unity Supply Co. — FOR Office supplies

BALANCE BROUGHT FORWARD 4564 00 — DEPOSITS — TOTAL — AMOUNT THIS CHECK 88 20 — BALANCE 4475 80

THOMAS B. ANDERSON
1234 MT. PLEASANT RD.
ANYWHERE, USA 12345
0103
64-60/611
June 12 19 XX
PAY TO THE ORDER OF Unity Supply Company $ 88.20
Eighty eight & 20/100 DOLLARS
Columbus Bank and Trust Company • Columbus, Georgia MEMBER FDIC
JOHN H. HARLAND COMPANY SAMPLE-VOID
FOR Office supplies Thomas B. Anderson
⑈000103⑈ ⑆061100606⑆ ⑈12 34 567⑈

JAN MARCANO, INTERIOR DECORATOR
1901 Ried Drive
Columbus, GA 31904

64-60
601

Date June 10 19 XX

Currency	153	00
Coin	2	35
Checks		
3-448	32	50
2-112	12	15
Total	200	00

Columbus Bank and Trust Company • Columbus, Georgia
MEMBER FDIC

⑆0601006666⑆ ⑈12 17 860⑈

Work Space

Name ____________________ Date ____________________

UNIT 53 Exercises

1. Fill in the balances in the following problem that shows the deposits and checks that were recorded on a series of check stubs.

Deposit	6,200.00
Check 1	401.98
BALANCE	
Check 2	304.88
BALANCE	
Check 3	677.11
BALANCE	
Deposit	975.80
BALANCE	
Check 4	1,245.66
BALANCE	
Check 5	433.55
BALANCE	
Deposit	3,458.95
BALANCE	
Check 6	567.89
BALANCE	

2. The check stubs below are from Bi-City Paint Company. The amount of each check and deposit has been recorded. Complete the stubs.

NO. 102 $ 10.50
DATE January 2 19 —
TO Kim Supply
FOR Supplies

	DOLLARS	CENTS
BALANCE	691	02
AMT. DEPOSITED	—	
TOTAL		
AMT. THIS CHECK	10	50
BALANCE		

NO. 103 $ 49.50
DATE January 3 19 —
TO Postal Service
FOR Stamps

	DOLLARS	CENTS
BALANCE		
AMT. DEPOSITED		
TOTAL		
AMT. THIS CHECK	49	50
BALANCE		

NO. 104 $ 418.00
DATE January 6 19 —
TO Central Power
FOR Power Bill

	DOLLARS	CENTS
BALANCE		
AMT. DEPOSITED	841	00
TOTAL		
AMT. THIS CHECK	418	00
BALANCE		

NO. 105 $ 67.80
DATE January 9 19 —
TO Dave Powers
FOR Lawn Care

	DOLLARS	CENTS
BALANCE		
AMT. DEPOSITED		
TOTAL		
AMT. THIS CHECK	67	80
BALANCE		

NO. 106 $
DATE 19
TO
FOR

VOID

	DOLLARS	CENTS
BALANCE		
AMT. DEPOSITED		
TOTAL		
AMT. THIS CHECK		
BALANCE		

NO. 107 $ 80.00
DATE January 15 19 —
TO Kim Supply
FOR Supplies

	DOLLARS	CENTS
BALANCE		
AMT. DEPOSITED	2140	00
TOTAL		
AMT. THIS CHECK	80	00
BALANCE		

NO. 108 $ 225.00
DATE January 19 19 —
TO Kathy Glove
FOR Salary

	DOLLARS	CENTS
BALANCE		
AMT. DEPOSITED	800	00
TOTAL		
AMT. THIS CHECK	225	00
BALANCE		

NO. 109 $ 195.00
DATE January 21 19 —
TO Head Service Co.
FOR Repairs

	DOLLARS	CENTS
BALANCE		
AMT. DEPOSITED		
TOTAL		
AMT. THIS CHECK	195	00
BALANCE		

NO. 110 $ 225.00
DATE January 30 19 —
TO Kathy Glove
FOR Salary

	DOLLARS	CENTS
BALANCE		
AMT. DEPOSITED		
TOTAL		
AMT. THIS CHECK	225	00
BALANCE		

Name ______________________ Date ______________________

UNIT 53 Applications

1. Using the check shown below, answer the following questions:

   a. Who is the payee?

   b. Who is the drawer?

   c. Who is the drawee?

   d. What is the amount of the check?

ANSWERS

1. a. __________
   b. __________
   c. __________
   d. __________

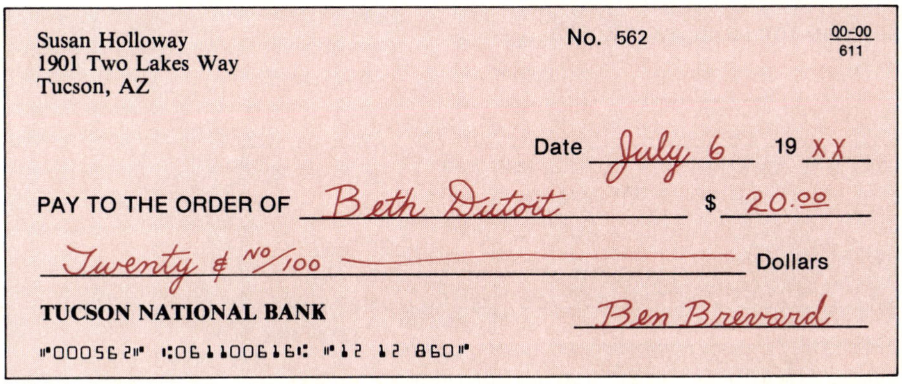

2. The deposits and cash expenditures of Case Equipment Company for June, 19XX, are shown below. Using the checkbook provided, write a check for each expenditure. Be sure to fill in the check stub before you write the check.

June 1, balanced forward, $6,210.58
2, wrote check #105 for $1,250 to Carter Realty for June rent.
8, made a deposit of $800.
8, wrote check #106 to Lot Service Co. for repairs, $238.50.
12, made a deposit of $2,680.
12, wrote check #107 to cash for salaries, $2,415.
18, made a deposit of $6,950.
18, wrote check #108 to Central Power Co., for utilities, $1,200.
26, wrote check #109 to cash for salaries, $2,415.

NO.______ $______
DATE__________19____
TO______________
FOR______________

	DOLLARS	CENTS
BALANCE		
AMT. DEPOSITED		
TOTAL		
AMT. THIS CHECK		
BALANCE		

NO.______
__________19____
PAY TO THE ORDER OF ______________________ $__________
______________________________________DOLLARS
____

NO.______ $______
DATE__________19____
TO______________
FOR______________

	DOLLARS	CENTS
BALANCE		
AMT. DEPOSITED		
TOTAL		
AMT. THIS CHECK		
BALANCE		

NO.______
__________19____
PAY TO THE ORDER OF ______________________ $__________
______________________________________DOLLARS
____

NO.______ $______
DATE__________19____
TO______________
FOR______________

	DOLLARS	CENTS
BALANCE		
AMT. DEPOSITED		
TOTAL		
AMT. THIS CHECK		
BALANCE		

NO.______
__________19____
PAY TO THE ORDER OF ______________________ $__________
______________________________________DOLLARS
____

Name ____________________ Date ____________________

NO.________ $__________
DATE __________ 19____
TO ____________________
FOR ____________________

	DOLLARS	CENTS
BALANCE		
AMT. DEPOSITED		
TOTAL		
AMT. THIS CHECK		
BALANCE		

NO. ________
__________ 19____
PAY
TO THE
ORDER OF ____________________ $__________
____________________ DOLLARS
____

NO.________ $__________
DATE __________ 19____
TO ____________________
FOR ____________________

	DOLLARS	CENTS
BALANCE		
AMT. DEPOSITED		
TOTAL		
AMT. THIS CHECK		
BALANCE		

NO. ________
__________ 19____
PAY
TO THE
ORDER OF ____________________ $__________
____________________ DOLLARS
____

NO.________ $__________
DATE __________ 19____
TO ____________________
FOR ____________________

	DOLLARS	CENTS
BALANCE		
AMT. DEPOSITED		
TOTAL		
AMT. THIS CHECK		
BALANCE		

NO. ________
__________ 19____
PAY
TO THE
ORDER OF ____________________ $__________
____________________ DOLLARS
____

Work Space

UNIT 54

Bank Reconciliation

THE BANK STATEMENT

A **bank statement** is a report that gives the status of a depositor's account according to the bank's records. The bank statement is prepared periodically, usually once a month. The statement lists the amounts deposited and withdrawn, interest earned on the account, if any, charges made by the bank against the account, and the balance of the account as of the date of the statement. Included with the bank statement will be all **canceled checks,** that is, checks that have cleared the account for payment during the period covered by the statement.

The following bank statement was received by Jan Marcano on July 3, 19XX.

RECONCILING THE BANK STATEMENT

The bank statement, as stated above, is a statement of the depositor's account according to the bank's records. And the checkbook balance is the depositor's record of the account. Thus, it would seem that the two records should agree in amount. But, in practice, this is rarely the case. This is not because of mistakes or inattention on the part of the depositor or the bank but, instead, is a result of several factors. Let's look at the common causes of disagreement between the bank statement balance and the checkbook balance.

1. **Deposits in transit.** Deposits in transit, also called **late deposits,** are deposits

TO: Jan Marcano
1901 Ried Drive
Columbus, GA 31904

Acct. No. 12 17 860

Columbus Bank & Trust
Columbus, GA

June 30, 19XX

Checks	**Deposits**	**Date**	**Balance**-$4,200
$200.00		6-2-XX	$5,120.00
125.00		6-4-XX	4,995.00
301.00		6-7-XX	4,694.00
109.00		6-9-XX	4,585.00
221.00	$200.00	6-10-XX	4,564.00
88.20		6-12-XX	4,475.80
322.80		6-18-XX	4,153.00
	1,400.00	6-22-XX	5,553.00
920.00		6-25-XX	4,633.00
856.10		6-28-XX	3,776.90
18.00 SC			
15.00 IP			3,743.90
	456.10	6-28-XX	4,200.00

Legend
DM = debit memorandum
CM = credit memorandum
SC = service charge
IP = imprinting checks
NSF = nonsufficient funds

Bank Statement

that did not reach the bank's accounting department in time to be added to the current bank statement. They appear in the checkbook, but not on the bank statement.

2. **Outstanding checks.** When a check is written by the depositor, the amount of the check should immediately be entered in the checkbook as a reduction in the checkbook balance. Once written, however, a check could take several days before it reaches the bank for payment. If the bank statement is prepared before a check clears the account for payment, the check will not show up on the bank statement. The check, then, is said to be outstanding.
3. **Service charges.** Banks usually charge a fee for providing checking accounts. This fee, called a service charge, is subtracted directly from the depositor's balance and is entered on the bank statement.
4. **Other bank charges.** There are certain other charges that banks make on a regular basis. These include charges for printing the checks, and charges for insufficient funds checks (ISF), that is, checks written for an amount greater than the balance of the account.
5. **Errors.** Errors are a common cause of disagreement between the checkbook balance and the bank statement balance. Most are made by the depositor, because banks today use electronic equipment for processing bank transactions.

When the bank statement balance and the checkbook balance do not agree, they must be reconciled, that is, brought into agreement. To reconcile the statement, follow these steps:

1. Add the amount of deposits in transit to the bank statement.
2. Subtract the amount of outstanding checks from the bank statement balance.
3. Subtract all charges appearing on the bank statement from the checkbook balance.

To illustrate the bank reconciliation process, let's look again at Jan Marcano's bank statement on page 465. According to the bank statement, Jan's account has a balance of $4,200. But, on the same date, Jan's checkbook shows a balance of $3,831. The two records were brought into agreement as follows:

1. By comparing each deposit recorded in her checkbook with the deposits appearing on the bank statement, Jan found that a deposit of $600 made on July 2 had not reached the bank's accounting department in time to be entered on the statement. So, this deposit is outstanding.
2. By arranging her canceled checks in numeric order and comparing each check recorded in the checkbook with those appearing on the bank statement, Jan found that several checks had not cleared her account. Specifically, the following checks were outstanding:

Check No.	Amount
221	$ 75
222	125
225	102
226	700

3. Jan checked the bank statement for any charge that may have been made against her account. She discovered that the bank had charged her account $18 for a service charge and $15 for printing checks.

Based on this analysis, Jan prepared the reconciliation statement shown below.

Jan Marcano, Interior Decorator
Bank Reconciliation
July 3, 19XX

Balance per bank statement		$4,200
Add deposit in transit		600
		$4,800
Deduct outstanding checks:		
#221	$ 75	
#222	125	
#225	102	
#226	700	1,002
Adjusted bank balance		$3,798
Balance per checkbook		$3,831
Deduct:		
Service charge	$ 18	
Imprinting checks	15	33
Adjusted checkbook balance		$3,798

Bank Reconciliation Statement

Name ______________________________ Date ______________________________

UNIT 54 Exercises

1. When reconciling a bank statement, state whether each of the following should be (a) added to the bank statement balance, (b) subtracted from the bank statement balance, (c) added to the checkbook balance, or (d) subtracted from the checkbook balance.

   a. ______ Outstanding checks

   b. ______ Deposits in transit

   c. ______ Service charge

   d. ______ Fee for printing checks

   e. ______ ISF check charge

2. List several factors that may prevent the bank statement balance from agreeing with the checkbook balance.

3. List, in order, the steps involved in reconciling a bank statement.

Work Space

Name ______________________ Date ______________________

UNIT 54 Applications

1. William Reynolds received his bank statement on June 30, 19XX. According to his statement, William has a balance of $2,400 in his checking account. According to William's checkbook, however, his balance is $2,788. Upon closer observation, William discovered the following:

   (a) An $800 deposit had not reached the bank in time to be entered in William's account when the statement was prepared.

   (b) The following checks were outstanding:
   #144 for $200
   #145 for 25
   #147 for 180
   #148 for 15

   (c) A service charge of $8 was made against William's account by the bank.

   Prepare a bank reconciliation statement in good form.

2. Reconcile the bank statements in the following problems.

(a)

Balance shown on the bank statement of Jill House, $655.45; balance as shown in the checkbook, $518.90.

Checks outstanding:

#135	$18.50
#138	22.80
#142	51.40
#145	19.02
#147	32.48

Service charge, $7.65.

(b)

Balance shown on the bank statement of Tri-City Paint Company, March 31, 19XX, $4,280.88; balance as shown in the checkbook, $4,035.13.

Deposit in transit, $312.00.

Outstanding checks:

#244	$125.00
#245	75.60
#248	34.89
#250	325.00
#253	19.21

Charge for printing checks, $12.50.

Service charge, $9.45.

APPENDIX A

Working with Measurements

Some business math problems will call for you to use various types of measurements. Measures of length, weight, surface, liquid, volume, time, and dozens are often used. The following table shows the measures that are most commonly used in business.

Problems involving measures generally are of three types:

1. Converting from one unit of measure to another.
2. Adding, subtracting, multiplying, or dividing measures.
3. Solving problems through the application of measures.

CONVERTING FROM ONE UNIT OF MEASURE TO ANOTHER

Use Table A-1 to help you when converting from one unit of measure to another. To convert a smaller unit of measure into a larger unit, you must divide.

TABLE A-1—MEASUREMENTS

Weight	
16 ounces (oz.)	= 1 pound (lb.)
2,000 pounds	= 1 ton
Length (Linear)	
12 inches (in.)	= 1 foot (ft.)
3 feet	= 1 yard (yd.)
5,280 feet	= 1 mile (mi.)
1,760 yards	= 1 mile
Time	
60 seconds (sec.)	= 1 minute (min.)
60 minutes	= 1 hour (hr.)
24 hours	= 1 day (da.)
7 days	= 1 week (wk.)
4 1/3 weeks	= 1 month (mo.)
52 weeks	= 1 year
12 months	= 1 year

Capacity	
2 cups	= 1 pint (pt.)
2 pints	= 1 quart (qt.)
4 quarts	= 1 gallon (gal.)
16 ounces (1 lb.)	= 1 pint
Area (Surface)	
144 square inches	= 1 square foot (sq. ft.)
9 square feet	= 1 square yard (sq. yd.)
43,560 square feet	= 1 acre (a.)
640 acres	= 1 square mile (sq. mi.)
Volume (Cubic)	
1,728 cubic inches	= 1 cubic foot (cu. ft.)
27 cubic feet	= 1 cubic yard (cu. yd.)
1 cubic foot	= 7½ gallons of water
Other Measures	
12 units	= 1 dozen (doz.)
12 dozen	= 1 gross (gr.)
144 units	= 1 gross

EXAMPLE 1: How many feet are in 168 inches?
Solution:

12 in. = 1 ft.

168 ÷ 12 = 14 feet

EXAMPLE 2: A large steak weighs 56 ounces. What is the weight of the steak in pounds?
Solution:

16 oz. = 1 lb.

56 ÷ 16 = 3.5 lbs. (or 3 lbs. and 8 oz.)

To convert a larger unit of measure to a smaller unit, it is necessary to multiply.

EXAMPLE 1: How many ounces are in 6 1/4 pounds?
Solution:

1 lb. = 16 oz.

$$6\tfrac{1}{4} \times 16 = \frac{25}{\overset{}{\underset{1}{\cancel{4}}}} \times \frac{\overset{4}{\cancel{16}}}{1} = 100 \text{ oz.}$$

EXAMPLE 2: Ben Taylor weights 190 lbs. What is his weight in ounces?
Solution:

190	lbs.
× 16	ounces in 1 lb.
3,040	ounces in 190 lbs.

EXAMPLE 3: How many quarts are in 12 gallons?
Solution:

1 gal. = 4 qt.

12 gal. × 4 = 48 qt.

ADDING, SUBTRACTING, MULTIPLYING, AND DIVIDING MEASURES

Addition. To add measures, you add the similar units separately, and then convert the excess units.

EXAMPLE 1: Add: 6 yards, 1 foot, 3 inches + 4 yards, 2 feet, 7 inches.
Solution:

	Yards	Feet	Inches	
	6	1	3	
	4	2	7	Add similar units together.
	—	—	—	
	10	3	10	
+	1	− 3		Convert excess units: 3 ft. = 1 yd.
	11		10	

EXAMPLE 2: Add 7 yards, 1 foot, 10 inches + 8 yards, 2 feet, 9 inches.
Solution:

	Yards	Feet	Inches	
	7	1	10	Add similar units.
	8	2	9	
	15	3	19	
		+ 1	− 12	Convert excess in. to ft.
	15	4	7	
+	1	− 3		Convert excess ft. to yd.
	16	1	7	

Subtraction. When subtracting measures, as in addition, you subtract similar units separately.

EXAMPLE: Subtract: 3 hours, 15 minutes from 5 hours, 30 minutes.
Solution:

	Hours	Minutes
	5	30
−	3	15
	2	15

In some subtraction problems, you will have to borrow from the larger unit.

EXAMPLE: Subtract: 12 yards, 2 feet, 7 inches − 4 yards, 11 inches.
Solution:

	Yards	Feet	Inches	
		1	19	
	12	~~2~~	~~7~~	You cannot subtract 11 in. from 7 in; so, you borrow 12 in. (1 foot) from the 2 feet: 7 in. + 12 in. = 19 in.
−	4		11	
	8	1	8	

Multiplication. To multiply units of measure, follow these steps:

Step 1: Convert the units being multiplied to one common unit.

Step 2: Multiply as usual.

EXAMPLE: Multiply 7 ft. 6 in. by 5 ft. 4 in.
Solution:

Step 1: Convert to one common unit:

7 ft. 6 in. = 7 6/12 ft. = 7 1/2 ft.
5 ft. 4 in. = 5 4/12 ft. = 5 1/3 ft.

12 in. = 1 ft.; thus, 6 in. = 6/12 ft.

Step 2: Multiply:

$$7\tfrac{1}{2} \times 5\tfrac{1}{3} = \frac{\overset{5}{\cancel{15}}}{\underset{1}{\cancel{2}}} \times \frac{\overset{8}{\cancel{16}}}{\underset{1}{\cancel{3}}} = 40 \text{ sq. ft.}$$

Division. To divide units of measure, you convert the units to one common unit and divide in the normal way.

EXAMPLE: Divide 6 gal. 3 qt. by 3
Solution:

Convert to one common unit:

6 gal. 3 qt. = 6 3/4 gal.

1 qt. = 1/4 gal.; thus, 3 qt. = 3/4 gal.

Divide:

$$6\tfrac{3}{4} \div 3 = \frac{\overset{9}{\cancel{27}}}{4} \times \frac{1}{\underset{1}{\cancel{3}}} = \frac{9}{4} = 2\tfrac{1}{4} \text{ gal.}$$

SOLVING BUSINESS PROBLEMS INVOLVING MEASUREMENTS

Many everyday business problems call for the use of measurements. For example, the cost of carpet is quoted in terms of square yards. Let's assume that you wanted to carpet your den with carpet that is being advertised at "$12 per square yard, installed." Carpeting covers an area (or a surface). Thus, you will be working with square measures. The first thing you will need to do is measure the length and width of the room. Let's assume that the room is 24 feet long and 18 feet wide. Now, you obtain the square feet in your den by multiplying length by width:

24 × 18 = 432 sq. ft.

Now, look at Table A-1 and find that there are 9 square feet to the square yard. So, you divide 432 sq. ft. by 9 to obtain the number of square yards:

432 ÷ 9 = 48 sq. yds.

You now find the price of the carpet by multiplying the cost of the carpet by the number of square yards needed:

48 sq. yds. × $12 per sq. yd. = $576

Let's look at another example.

EXAMPLE: How many gallons will it take to fill a tank that is 20 feet long, 7 feet wide, and 4 feet high?
Solution:

Length	×	Width	×	Height	=	Volume
20	×	7	×	4		560 cu. ft.

560 cu. ft. × 7 1/2 gal. per cu. ft. = 4,200 gal.

Work Space

Name ______________________________ Date ______________________________

APPENDIX A Exercises

1. How many feet are there in:
   (a) 228 inches ______

   (b) 276 inches ______

   (c) 78 inches ______

2. How many ounces are there in:
   (a) 8 pounds ______

   (b) 16 pounds ______

   (c) 5 3/4 pounds ______

3. How many pints are in 12 gallons?

4. How many hours are in 5,400 seconds?

5. Convert the following:
   (a) 54 inches to feet ______

   (b) 36 feet to yards ______

   (c) 5,184 cubic inches to cubic feet ______

   (d) 63 square feet to square yards ______

(e) 3 miles to yards ______

(f) 128 ounces to pounds. ______

(g) 21,120 feet to miles ______

(h) 8 tons to pounds ______

(i) 3 acres to square feet ______

6. Add the following:
   (a) 6 yards, 2 feet, 7 inches + 5 yards, 1 foot, 11 inches

   (b) 12 yards, 9 inches + 3 yards, 2 feet, 10 inches + 8 yards, 1 foot

   (c) 6 hours, 45 minutes + 3 hours 15 minutes

Name ____________________ Date ____________________

(d) 4 hours, 20 minutes, 30 seconds + 2 hours, 40 minutes, 50 seconds

7. Subtract the following:
   (a) 6 hours, 18 minutes − 3 hours, 30 minutes

   (b) 12 hours, 12 minutes, 48 seconds − 4 hours, 10 minutes, 52 seconds.

   (c) 5 yards, 2 feet, 3 inches − 2 yards 1 foot, 8 inches

8. Perform the following:
   (a) Multiply 3 feet, 4 inches by 2 feet, 6 inches

   (b) Multiply 2 feet, 6 inches by 10 inches

(c) Divide 4 gallons, 2 quarts by 2

(d) Divide 6 feet, 8 inches by 5

Name ______________________ Date ______________________

APPENDIX A Applications

1. How many square yards of linoleum tile would it take to cover a kitchen that is 30 ft. wide and 36 ft. long?

2. If a 6 ft. 9 in. pipe is cut into 3 equal pieces, what would be the length of each piece?

3. How many gallons of water would it take to completely fill a tank that is 24 ft. by 8 ft. by 6 ft.?

4. Sue Taylor weighs 1,680 ounces. What is her weight in pounds?

5. A room 24 ft. by 18 ft. is to be carpeted. How much would it cost to cover the room if the carpeting is $8.00 per square yard, installation is $2.00 per square yard, and padding is $1.50 per square yard?

6. Alan can run a mile in 8 minutes and 18 seconds. Todd can run a mile in half Alan's time. What is Todd's time?

Work Space

APPENDIX B Compound Interest

In Unit 50, you learned that interest is the charge for credit. You also learned how to calculate interest using the formula I = PRT. To review this formula, let's assume that you borrowed $1,000 at a rate of 12% for 1 year. Your interest charge would be:

P	×	R	×	T		
$1,000	×	.12	×	1	=	$120

In this example, the $120 interest would be a charge to you because you had the use of someone else's money. But interest is a two-way street. When you borrow money, you pay interest; when you save money with a savings bank (or other savings institution), you earn interest. Interest earned by you is usually "compounded," a topic we will discuss next.

CALCULATING COMPOUND INTEREST

When you save with a savings bank, your money earns **compound interest.** This means that interest earned during each interest period is added to the principal before interest for the next period is calculated. Thus, the base on which interest is calculated becomes greater and greater.

To illustrate how compound interest works, let's assume that you deposit $500 in a bank account that is currently paying 8% interest. And let's further assume that you left your money on deposit with the bank for 2 years. Your compound interest for the two years is calculated as follows:

First year:

$500 × .08 × 1 = $40 interest

$500 + $40 = $540 compound amount

Second Year:

$540 × .08 × 1 = $43.20 interest

$540 + $43.20 = $583.20 compound amount

$583.20	compound amount
− 500.00	original principal
$ 83.20	compound interest

In this example, there were two interest periods, which gave us a compound interest calculation of $83.20. Compare this amount with interest calculated on the principal only (interest calculated on the principal only is called **simple interest**):

$500 × .08 × 2 = $80

As you can see, a simple interest calculation gives us $80 interest for two years, whereas, compound interest results in interest of $83.20. So, your money would earn $3.20 more ($83.20 − $80.00) simply by the way the interest is calculated.

CALCULATING COMPOUND INTEREST FOR SHORTER PERIODS

Interest is usually compounded more than once a year. For example, interest can be compounded monthly, quarterly (every 3 months), and semiannually (every six months). The more frequently the compounding, the more interest you earn—because the base grows faster. Let's see how much interest $500 would earn if interest at 8% is compounded semiannually (twice a year) for 2 years:

First interest period:

$500	× .08 × 1/2 year = $20 interest
+ 20	
$520	compound amount

Second interest period:

$520.00	× .08 × 1/2 year = $20.80 interest
+ 20.80	
$540.80	compound amount

Third interest period:

$540.80	× .08 × 1/2 year = $21.63 interest
+ 21.63	
$562.43	compound amount

Fourth interest period:

$562.43	× .08 × 1/2 year = $22.50 interest
+ 22.50	
$584.93	compound amount

Compound interest:

$584.93	compound amount
−500.00	original principal
$ 84.93	compound interest

As you can see, when interest was compounded semiannually at 8% for two years, a $500 savings balance will earn $84.93. Compare this amount with the $83.20 figure we obtained earlier when interest on the same $500 was compounded annually for two years at 8%. By compounding the interest twice a year (instead of once a year), an additional $1.73 ($84.93 − $83.20) of interest was earned. Remember, the more frequently interest is compounded, the more interest you will earn.

USING A COMPOUND INTEREST TABLE

Compounding interest for a large number of interest periods can be very time consuming. As a result, compound interest tables are often used.

Table B-1 shows compounded amounts for $1 at various interest rates for various time periods. To use the table, follow these steps:

Step 1: Place your finger on the correct number of interest periods in the far left column labeled "Number of Interest Periods."

Step 2: Find the appropriate interest rate at the top.

Step 3: Move across from the interest period and down from the interest rate until you meet; the number listed is the compound total for $1.

Step 4: Multiply the principal by the compound total for $1.

EXAMPLE 1: Use Table B-1 to calculate the compound interest on $600 for 5 years compounded annually at 8%.
Solution: Find period 5 in the first column and move across to the 8% column; you will find 1.46933. Multiply this number by the principal to find the compound amount:

$600 × 1.46933 = $881.60 compound amount

$881.60	compound amount
−600.00	principal
$281.60	compound interest

EXAMPLE 2: Use Table B-1 to calculate the compound interest on $3,000 for 10 years compounded annually at 12%.
Solution: Find period 10 in the first column and move across to the 12% column; you will find 3.10585. Multiply this number by the principal to find the compound amount:

TABLE B-1—COMPOUND INTEREST TABLE (Amount of $1 at Compound Interest)

Number of Interest Periods	Rate per Period								
	1%	2%	3%	4%	5%	6%	8%	10%	12%
1	1.01000	1.02000	1.03000	1.04000	1.05000	1.06000	1.08000	1.10000	1.12000
2	1.02010	1.04040	1.06090	1.08160	1.10250	1.12360	1.16640	1.21000	1.25440
3	1.03030	1.06121	1.09273	1.12486	1.15763	1.19102	1.25971	1.33100	1.40493
4	1.04060	1.08243	1.12551	1.16986	1.21551	1.26248	1.36049	1.46410	1.57352
5	1.05101	1.10408	1.15927	1.21665	1.27628	1.33823	1.46933	1.61051	1.76234
6	1.06152	1.12616	1.19405	1.26532	1.34010	1.41852	1.58687	1.77156	1.97382
7	1.07214	1.14869	1.22987	1.31593	1.40710	1.50363	1.71382	1.94872	2.21068
8	1.08286	1.17166	1.26677	1.36857	1.47746	1.59385	1.85093	2.14359	2.47596
9	1.09369	1.19509	1.30477	1.42331	1.55133	1.68948	1.99900	2.35795	2.77308
10	1.10462	1.21899	1.34392	1.48024	1.62889	1.79085	2.15892	2.59374	3.10585
11	1.11567	1.24337	1.38423	1.53945	1.71034	1.89830	2.33164	2.85312	3.47855
12	1.12683	1.26824	1.42576	1.60103	1.79586	2.01220	2.51817	3.13843	3.89598
13	1.13809	1.29361	1.46853	1.66507	1.88565	2.13293	2.71962	3.45227	4.36349
14	1.14947	1.31948	1.51259	1.73168	1.97993	2.26090	2.93719	3.79750	4.88711
15	1.16097	1.34587	1.55797	1.80094	2.07893	2.39656	3.17217	4.17725	5.47357
16	1.17258	1.37279	1.60471	1.87298	2.18287	2.54035	3.42594	4.59497	6.13039
17	1.18430	1.40024	1.65285	1.94790	2.29202	2.69277	3.70002	5.05447	6.86604
18	1.19615	1.42825	1.70243	2.02582	2.40662	2.85434	3.99602	5.55992	7.68997
19	1.20811	1.45681	1.75351	2.10685	2.52695	3.02560	4.31570	6.11591	8.61276
20	1.22019	1.48595	1.80611	2.19112	2.65330	3.20714	4.66096	6.72750	9.64629
21	1.23239	1.51567	1.86029	2.27877	2.78596	3.39956	5.03383	7.40025	10.80385
22	1.24472	1.54598	1.91610	2.36992	2.92526	3.60354	5.43654	8.14027	12.10031
23	1.25716	1.57690	1.97359	2.46472	3.07152	3.81975	5.87146	8.95430	13.55235
24	1.26973	1.60844	2.03279	2.56330	3.22510	4.04893	6.34118	9.84973	15.17863
25	1.28243	1.64061	2.09378	2.66584	3.38635	4.29187	6.84848	10.83471	17.00006
26	1.29526	1.67342	2.15659	2.77247	3.55567	4.54938	7.39635	11.91818	19.04007
27	1.30821	1.70689	2.22129	2.88337	3.73346	4.82235	7.98806	13.10999	21.32488
28	1.32129	1.74102	2.28793	2.99870	3.92013	5.11169	8.62711	14.42099	23.88387
29	1.33450	1.77584	2.35657	3.11865	4.11614	5.41839	9.31727	15.86309	26.74993
30	1.34785	1.81136	2.42726	3.24340	4.32194	5.74349	10.06266	17.44940	29.95992

$3,000 × 3.10585 = $9,317.55 compound amount

$9,317.55 − $3,000 = **$6,317.55 compound interest**

USING A TABLE FOR PERIODS OF LESS THAN A YEAR

In Examples 1 and 2 above, we used Table B-1 to calculate compound interest on an annual basis. When interest is compounded annually, the number of interest periods is the same as the number of years. But when you use Table B-1 to compound interest for shorter periods, you must adjust the interest periods and the annual rate. To do so, follow these steps:

Step 1: Multiply the number of years by the number of times interest will be compounded in the year.

Step 2: Divide the interest rate by the number of times interest will be compounded during the year.

Step 3: Using these adjusted figures from Steps 1 and 2, calculate the compound interest.

EXAMPLE 1: Use Table B-1 to calculate the compound interest on $300 for 3 years compounded quarterly at 8%.

Solution: Quarterly compounding = 4 times a year

Step 1: Multiply 3 years by 4:

3 years × 4 = 12 interest periods

Step 2: Divide the annual interest rate by 4:

$$8\% \div 4 = 2\%$$

Step 3: Find period 12 in the first column of Table B-1 and move across on the same line to the 2% column; you will find 1.26824. Multiply this figure by the principal to get the compound amount:

$300 × 1.26824 = $380.47 compound amount

$380.47	compound amount
−300.00	principal
$ 80.47	**compound interest**

EXAMPLE 2. Find the amount of compound interest earned on a $3,000 investment at 12% for 2 years if interest is compounded monthly.
Solution: Monthly compounding = 12 times a year

Step 1: 2 years × 12 = 24 interest periods

Step 2: 12% ÷ 12 = 1%

Step 3: Find 24 in the first column of Table B-1 and move across to the 1% column; you will find 1.26973. Multiply this figure by the principal:

$3,000 × 1.26973 = $3,809.19 compound amount

$3,809.19 −$3,000 = **$809.19 compound interest**

Name ______________________ Date ______________________

APPENDIX B Exercises

Without using Table B-1, find the compound interest on each of the following:

	Principal	Rate	Time	Compounding Period	Compound Interest
1.	$ 800	12%	1 year	quarterly	$_____
2.	$1,200	8%	3 years	annually	$_____

	Principal	Rate	Time	Compounding Period	Compound Interest
3.	$5,000	10%	5 years	semiannually	$_____
4.	$600	9%	2 years	quarterly	$_____

Name ______________________ Date ______________________

Use Table B-1 to complete the following:

	Principal	Rate	Time	Compounding Period	Compound Interest
5.	$ 4,500	6%	3 years	annually	$_____
6.	$12,000	10%	5 years	semiannually	$_____
7.	$ 6,000	8%	4 years	quarterly	$_____
8.	$20,000	12%	2 years	monthly	$_____

Work Space

Name ________________________ Date ________________________

APPENDIX B Applications

1. Lyle Burgman deposited $700 in his savings account two years ago. His bank is paying 8% interest compounded quarterly. How much interest has he earned?

2. Julie Gamble deposited $2,000 in her savings account five years ago. What is the balance of her account today if she earned 12% interest compounded quarterly?

3. What interest would a $12,000 savings balance earn in one year if the bank is paying 7% interest compounded semiannually?

4. How much interest would $6,000 earn in 2 years if it is deposited in an account paying 12% interest compounded monthly?

Work Space